WAS EINSTEIN RIGHT?

NOT QUITE! (TM)

'Perfecting $E=mc^2$' And

'Debunking Quantum Weirdness'

Now Leads To

'The Holy Grail of *Unified Physicality*'!

(Preview Monograph Edition)

By: N. Glenn Gratke, Ph.D.

Legal Notice Page

Cover Artwork: Title: ***"A Great Spirit Ponders"***

- Composed and Commissioned by Norman Glenn Gratke
- Artfully Drawn by Shawn Tuft of Waukesha, Wisconsin
- Copyright Registered October 2005 by Norman Glenn Gratke

Book Web Cite: For sales and promotion information, visit:

- **Emc2GG.com**
- **emc2gg.com**

Author Email: To contact the author, send email to:

- **DrG@Emc2GG.com**
- **drg@emc2gg.com**

ISBN: 1-4196-1628-5

Preamble

Eureka (I believe I have found it),
'The Holy Grail of Unified Physicality'!

First and foremost, this *Monograph* is an original and revolutionary scientific treatise. I envision a '*Unified Physical Field Theory*' that unifies the description of all natural phenomena under the improbable banner of *Common Sense Physicality.* I feel compelled to expunge the mystical interpretations currently in vogue within the Relativity, the Quantum Mechanics, and the Particle Theories of Modern Physics. I thus seek to unify physics within a context of *Objective Physical Reality.*

The Quantum Mechanics of Modern Physics interprets the idea of 'physical existence' in a very strange and distorted way. 'Supposedly', an object dissolves when no one looks at it, and then it re-materializes when just one person sneaks a peek at it. This 'thought-controlled existence' is utterly fantastic and most absurd. Einstein rejected this 'way of thinking', and said,

"Quantum Mechanics is very worthy of regard.
But an inner voice tells me that
this is not yet the right track.
The theory yields much,
but it hardly brings us closer
to the Old One's secrets."

Einstein thought objects should be able to exist independent of our thoughts about them, because objects should possess *Objective Physical Reality.* On this point, I strongly agree with Einstein and will endeavor herein to vindicate this view.

I contend that the mathematical presentations of Modern Physics are flawed in some very subtle ways. By slightly changing the mathematical formulations, we can completely alter the results. We can still explain the observations properly attributed to today's physics, but we can then restructure Modern Physics to conform to a viable vision of *Common Sense Physicality.*

We (you the reader and I) shall consider the essence of space, time, elementary particles, and elementary forces. We shall revise our perceptions of basic structures, from tiny atoms up to enormous galaxies. With my vision of *Objective Physical Reality*, I propose a radical 'paradigm shift'. I propose to solve the greatest mystery in all of Physics and unveil the discovery of 'The Holy Grail of *Unified Physicality*'.

Albert Einstein's vision of a 'Unified Field Theory' has greatly inspired me in my task. I would thus like to honor Albert Einstein during the year of 2005, which is the centennial of his miracle year of 1905. The highest honor I can bestow on him is to proclaim he was essentially correct in his grand vision. However, we should also respect his humanity and be prepared to acknowledge the errors and omissions that prevented his realization of *Ultimate Unification*.

Without wavering, I seek to expose the fallacies that improperly ensnare the most basic tenets of Modern Physics. Herein I seek to outline the completion of Einstein's work. This is *"The best book ever written about Einstein's ideas"*, because it alone elevates his abstractions to the perception of *Objective Physicality*. This is *"One of the best books ever written about Physics"*, because it alone breaks the intellectual barriers that have been blocking fulfillment of *Ultimate Unification*.

I believe my vision of the '*Unified Physical Field Theory*' shall become a great milestone in the history of science. I believe this scientific revelation will unveil the discovery of 'The Holy Grail of *Unified Physicality*'. Will you be able to read with an unbiased and open mind? If so, you too may come to accept my vision of *Unified Physicality*!

About the Author

Ryan,
Good luck in your Studies.
Beware of falsehoods masquerading as truth.
Be smart and expose them if you can.
Dr. G.

N. Glenn Gratke was born May 10, 1945 in Milwaukee Wisconsin. He was the third of six children born to Paul and Conice Gratke. He attended public schools in Milwaukee, and as a young man, went on to earn a Bachelor's Degree in Electrical Engineering from the University of Wisconsin in 1967.

In 1967, Vietnam era military service requirements prevented his immediate pursuit of graduate studies. He served six years in the United States Air Force, and rose to the rank of Captain. He served all that time in a technical capacity as an Satellite Systems Engineer assigned to the Vela Satellite Program. Thereafter, he returned to Wisconsin and worked in industry as an Electrical Engineer. His primary area of expertise involved computers, computerized controls, and computer software. During that time he also undertook graduate courses in Electrical Engineering at the University of Wisconsin in Milwaukee.

In 1989, he made a major change-of-life decision. He discovered some key insights about Einstein's goal of unification. He decided to pursue graduate studies in Physics at UWM, and subsequently earned both a Masters and a Doctoral degree in Physics. He now is an Associate Professor of Physics at the Milwaukee School of Engineering (MSOE).

In this *Monograph*, Glenn Gratke presents the results of his efforts over the last 16 years to develop his ideas about *Unified Physicality.*

Table of Contents

Note on Editions & Organization

The *Full-Book Edition* will be presented in five parts, where *Part I* is being pre-released as this stand-alone *Monograph*. In the future, both of these editions will be re-released in more fully *Illustrated Versions*. The initial objective is to make the fantastic scientific revelation of *Unified Physicality* available to both physicists and the general public. The main point to realize is that this story of *Unified Physicality* drastically changes the status quo of physics, so timeliness in the release of information is essential.

Part I - 'Introducing Unified Physicality': introduces the topic of the *Unified Physical Field Theory*. A *Personal Preface, 'About the Title, Topic, and Me'*, provides an autobiographical reflection for developing the title, subtitle, and cover artwork of this monograph along with my vision for the main topic of *Unified Physicality*. A *Prologue, 'Building Upon Einstein's Legacy'* pays special tribute to Albert Einstein for his inspiring dream of a 'Unified Field Theory'. More than any other physicist who ever lived, Einstein has inspired me to seek this vision of *Unified Physicality*. *Chapter 1, 'The Vision of Unified Physicality',* introduces and overviews the key issues of *Physical Unification*. I especially urge everyone to carefully read this first chapter. Thereafter, in the *Full-Book Edition*, feel free to read whatever appeals to you in whatever order you like. Your subsequent reading can then proceed with balanced perspective and hopeful anticipation.

Part II - 'Perspectives About Unification': (only in the *Full-Book Edition*) shall provide a philosophical and historical perspective of physics. I will confront the 'sophistry of anti-physicality'. Then, I will review the ideas of the scientific pioneers of ancient times, of Newton's Era of Mechanics, of Maxwell's Era of Electromagnetism, and of Einstein's Era of Modern Physics. The past ideas of physics are a wonderful backdrop for a curious mind.

Part III - 'The Fallacies of Modern Physics': (only in the *Full-Book Edition*) shall detail the 'anti-physicalities' and other errors within Modern Physics. I will systematically expose the fallacies of Special Relativity, General Relativity, Quantum Mechanics, and Particle Theories. Modern Physics has been improperly built upon a foundation that is the very antithesis of *Objective Common Sense Physicality.*

Part IV - 'The Unified Physical Field Theory': (only in the *Full-Book Edition*) shall present my vision of *Unified Physicality* along with some significant ramifications for understanding what we observe in our universe. I will explain more fully the field formation of particles and force fields. I will more carefully describe the *Cosmic Continuum*, *Field Strings*, and *Cosmutons.* I will more precisely describe the enhanced role of Electromagnetism in the buildup of matter from tiny atomic nuclei up to enormous galaxies. An *Epilogue* will then summarize and reflect upon this subject of *Physical Unification*, and then go on to speculate about the origins of the universe and misguided intellectualism.

Part V - 'Supplemental Stuff': (only in the *Full-Book Edition*) shall include a *Glossary*, some *Appendices*, and an *Index*. Hopefully this should to provide additional utility to the *Full-Book Edition.*

Dedications

The world will be a better place
when we all learn to live
up to the responsible ideal of family.

In general, I dedicate this work to the ideal of family, the most trustworthy building block of civilized society. Why dedicate a book about physics to the family? Because physics is an idealistic adventure in science, and the family is an idealistic structure within civilization. The idealism of physics and the idealism of the family have now become the two center pieces in my life.

In particular, I dedicate this work to my own immediate family.

My parents (both deceased), Paul and Conice.

My most precious wife, Julienne.

Our wonderful children, Karl and Sarah Lynn.

Their respective spouses, Sarah Ann and Mike.

Our newest joy, our grandson Kade.

Any future grandchildren.

In addition, I dedicate this work to all those who have ever yearned and labored for a vision of a *Unified Common Sense Physicality*. I salute their spirit and their efforts.

Acknowledgements

What does it take to go from being a 44-year-old career Electrical Engineer in 1989 to being a 60-year-old University Professor of Physics in 2005 and formally propose a *Unified Physical Field Theory*? It takes some good ideas, a lot of work, a lot of patience, and a lot of support. My family has been very supportive and to them (particularly my deceased mother Conice, and most especially my wife Julienne) I owe special acknowledgement.

Since I did not fully disclose the intent of my proceeding 16 years of research to anyone outside my immediate family, outsiders could not have known the full scope of my work. To those who have known me in this period, I ask your forgiveness for not fully informing you of my activities. To those who have assisted (sometimes without even realizing it), I thank you for your kindness. While there are many people who have assisted in one-way or another, there are several people who I wish to mention by name.

Over the last 16 years, I have split my time one-way or another between two universities. These are the University of Wisconsin in Milwaukee (UWM) where I earned my masters and doctoral degrees in physics, and the Milwaukee School of Engineering (MSOE) where I now a professor teaching physics. Here are the most notable of individuals at UWM and MSOE who facilitated my efforts in some important way.

Professor Emeritus Dale Snider, UWM Department of Physics: Dale was my Ph.D. advisor. I owe him a great deal for helping to guide me through the graduate program. Dale was a unique resource of diverse knowledge and a champion of 'Chaos Theory'. It was his insights about 'Chaos' that initially attracted me to him. My interest in 'Chaos' was that its hidden dynamics should describe the time evolution and statistical randomness of quantum state transitions. However, for my Ph. D. thesis I undertook "*Quantum Ray Mapping*", a semi-classical way to map the evolution of quantum waves for an electron beam being scattered by gaseous atoms. We were both pleasantly surprised with the success of my research to 'Map Quantum Waves'. Mostly though, Dale helped me to better understand some of the formalities of 'Chaos' and Quantum Mechanics.

Professor John Friedman, UWM Department of Physics: John was my first choice as a Ph. D. advisor. Though we could not come to settle on a topic, John spent time helping me to better understand the formalities of Relativity. While advisors try to guide students into their own areas of expertise, I tried to guide John into my area of interest. When John reads this *Monograph* and the subsequent *Full-Book Edition*, he will see some of the very ideas I suggested to him in the early 1990s.

Professor Joe McPherson (deceased), UWM Department of Electrical Engineering: Joe and I went back to the middle 1960s. When UWM offered its first classes in Electrical Engineering, Joe (as a young energetic Ph. D.) arrived to inspire the new Electrical Engineering students. In the early 1970s, when the Air Force released me from my Vietnam Era service, I returned to civilian life and renewed my acquaintance with Joe as I undertook graduate studies in Electrical Engineering. In 1989, when I began graduate studies in Physics, I again renewed contact with Joe. Joe was a friend and advisor, and we talked about many things. In 1994, Joe suggested and helped me attain my initial teaching appointment at MSOE. Joe died much too young, and I still miss his infectious laugh.

Professor Anders Schenstrom, MSOE Department Chairman of Physics and Chemistry: Anders made available essential computing resources, which enabled me to complete my Ph. D. research. He also granted me valuable academic latitude. Coming from industry, I was used to some latitudes, but within carefully defined boundaries. In academia, I have been pleased to experience wider boundaries. In particular, Anders granted me the opportunities to develop and teach two new courses. The courses "*Relativity and Cosmology*" and "*Topics in Unification*" have allowed me to keep key ideas in critical focus.

Gary Shimek, MSOE Library Director: Gary has been a life preserver for my research. MSOE is a small, private institution, but Gary turns a small college library into a large tool for research. He and his staff obtained many references from diverse locations. In addition, his thoughtful reading and reviewing of my early drafts of this work has been very helpful, though necessarily brutal.

My wife Julienne has been my most valued resource for editing my work. I could not have written this book without her. She edits with a red pen that can boldly slice out needless wordiness like a mighty sword, or delicately dissect nuanced thoughts like a surgeons scalpel.

Among all others who volunteered to serve as readers for any of the various drafts of this work, one individual stands out. Jake Maciejewski (a recent MSOE Graduate in Electrical Engineering) has been the most diligent, the most timely, the most complete, and the most relevant with his comments. Several other readers reading various parts and drafts of this work have also contributed to its final form. I really appreciate all readers for their assistance. Of course, the responsibility for any errors and bad form still reside with me and not any reader.

Many MSOE students have attended my courses. Their enthusiasm and curiosity continuously rejuvenates me and helps to make teaching a rewarding experience. I especially appreciate the thoughtful questions of the more astute students.

I would also like to acknowledge all those who have yearned and labored for a vision of a *Unified Common Sense Physicality*. In this group I include not just 'respected scientists', but also 'scientific heretics', who are usually ostracized for their unorthodoxies. You may wonder why I would dare associate with unorthodox fellows. Why do criminals rob banks? Because the banks contain money. Why did Jesus associate with tax collectors and sinners? Because they needed ministerial attention. Why do I consider the ideas of 'alternate theory proponents' and the discarded ideas of 'respected scientists'? Because that is where some prematurely rejected truths are hiding and await our re-discovery.

I would like to acknowledge a few 'alternate theory proponents' for their audacious boldness of their thinking and their sowing of some complementing seeds of thought. Immanuel Velikovsky (an electromagnetic universe), Halton Arp (quasar/galaxy proximity), and David Bergman (electron ring model) are three of my favorite 'alternative theory proponents'. Of course, it is not proper to seriously cite such unorthodox fellows, because their ill repute will then be attributed to the one citing them. Fallacious 'guilt-by-association' is shamefully small-minded. New ideas require 'thinking outside the box', which sounds risky to those who fearfully dwell within the box and who seek to avoid 'rocking the boat'.

Among the 'most respected scientists', I would like to acknowledge Sir Isaac Newton (density-gradient gravitational aether), James Clerk Maxwell (thermal-luminiferous electromagnetic aether), and Albert Einstein (the 'Total Field' aether). Their disrespected visions of *Objective Physical Reality* are actually quite worthy and deserve serious consideration. I combine and resolve their visions of 'aether' into my own vision of the '*Cosmic Continuum*'.

Part I

Introducing Unified Physicality

Let us Contemplate

'The Holy Grail of Unified Physicality'.

How can I begin to tell you this amazing story about *Unified Physicality*? In struggling greatly with this question, I have finally settled on the following approach. First, within a *Personal Preface*, let me tell you how I became involved. Second, within a *Prologue*, let me then tell you how Albert Einstein relates to this story of unification. Then third, within *Chapter 1*, let me formally introduce you to a specific overview of my 'Vision of Unified Physicality'. The subsequent *Parts* of the *Full-Book Edition* will embellish upon this *Part I* introduction, which is being released herein as a stand-alone *Monograph*.

"A Great Spirit Ponders"

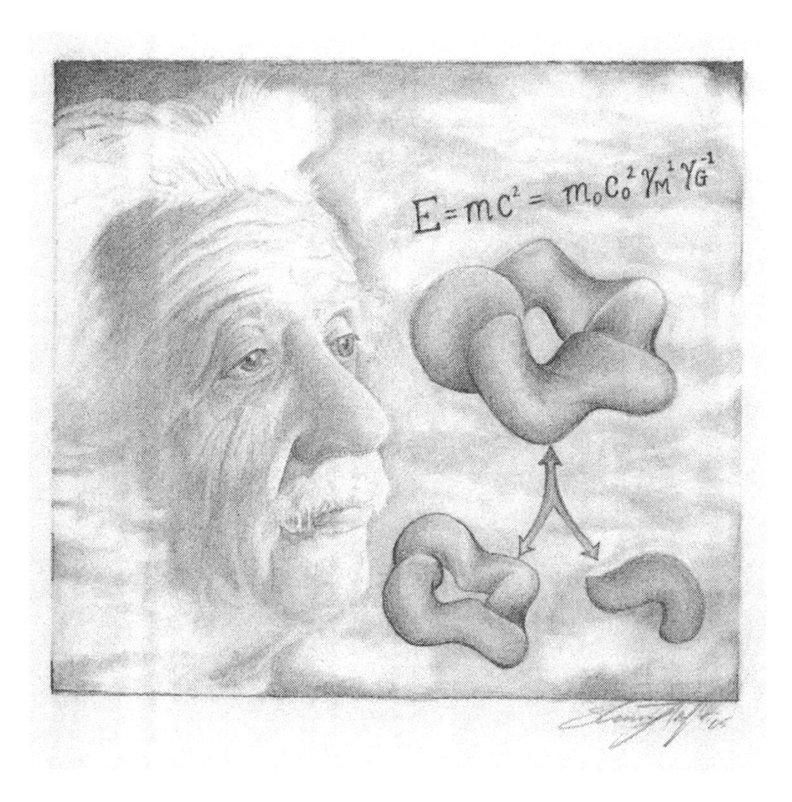

The 'Great Spirit' of Albert Einstein is depicted pondering my vision of $E=mc^2$ and a mutating *Cosmuton*. (This commissioned artwork has been drawn by Shawn Tuft of Waukesha, Wisconsin.)

Personal Preface

About the Title, Topic, and Me

Prepare to behold a Scientific Revelation: 'A Theory of Unified Physicality'!

Maybe the word 'Theory' is too strong for the yet to be convinced reader, and some may prefer the word 'Hypothesis'. Only when the world community of physicists digests this 'Hypothesis of *Unified Physicality*' and grants its stamp of approval can this be accepted as a 'Theory of *Unified Physicality*'. Some might argue that even 'Hypothesis' is too presumptuous. At this stage of introduction, if the word 'Theory' or 'Hypothesis' seems too much, I invite you to consider my 'Vision of *Unified Physicality*'.

Personal Preface - Article 1

A Major Paradigm Shift

Whichever tone you prefer (theory, hypothesis, or vision); I ask your indulgence to put yourself into a positive frame of mind. Read with the presumption that this vision is a '*fait accompli*'. I thus ask that you read as if the vision is complete, even though future refinements will develop a more polished rendition. I perceive this 'Vision of *Unified Physicality'* to be a major 'paradigm shift' and to constitute a new beginning for physics.

Science typically evolves with steady progress that builds upon prior accomplishments. A continuity of evolving thought develops as progressive refinements solidify around previously enunciated basic principles. However, every once in a while, something major happens to disrupt this slow and steady evolution. We discover a major change of basic principles. Such changes are known as 'paradigm shifts'. Initially accepting such changes can be difficult for many, because it upsets the 'way of thinking' that has become quite familiar.

History has its precedents. The Copernicus-Newton Revolution in the 1500s and 1600s inaugurated the era of Classical Mechanics. Newton unified terrestrial (on the earth) mechanics with celestial (above the earth) mechanics. The Faraday-Maxwell Revolution in the 1800s created the era of Electromagnetism. Maxwell unified electricity, magnetism, light, and thermal radiation under the banner of Electromagnetism. The Planck-Einstein Revolution of the early 1900s launched us into the era of Modern Physics but with conflicting visions for Relativity and Quantum Mechanics. Einstein unveiled several wondrous ideas beginning just 100 years ago which we will discuss later in the *Prologue*. Einstein then sought to unify Newton's Mechanics and Maxwell's Electromagnetism with the Relativity and Quantum Mechanics of Modern Physics. However, he was not able to solve this puzzle, and thus, he was unable to complete his grand vision of unification.

I seek to unify the revelations of Copernicus, Newton, Faraday, Maxwell, Planck, Einstein and others to achieve Einstein's great dream of *Ultimate Unification*. Should my revelation be valid, it would most assuredly be a major 'paradigm shift'. Once physicists feel properly justified to accept this proposal, they will resume their steady progress. Subsequently, they will then develop progressive refinements that solidify around the new basic principles underlying '*Unified Physicality*'.

I herein offer my vision for a '*Unified Physical Field Theory'*. My work to develop and justify this vision of *Unified*

Physicality is not complete, but I cannot delay any longer to unveil this scientific revelation. I have already delayed too long in openly disclosing my prior discoveries. As you consider my proposition, I ask you to envision the overall scope and merit of what I propose and not let any possible flawed detail unnecessarily dissuade you. It is for trying to reach perfection and to avoid foolish blunders that I have delayed so long, but a completely refined description will have to wait. I invite anyone, who might feel so inspired, to offer constructive suggestions to improve upon my vision for a '*Unified Physical Field Theory*'.

In the balance of this *Personal Preface*, let me provide an autobiographical reflection for developing the title, subtitle, and cover artwork of this *Monograph*. This in itself goes a long way in explaining my vision for 'The Holy Grail of *Unified Physicality*'.

Personal Preface - Article 2

"WAS EINSTEIN RIGHT? NOT QUITE!"

In 1989, being 44 years old with an on-going 23-year career as an Electrical Engineer, I undertook a mid-life redirection. Early in that year, my wife's brother (Ralph Evan - just a year younger than me) unexpectedly passed away. His passing stimulated me to reflect on the course of my own life. What is important in life? I had not done all that I had wanted to do, and I was not achieving my most personal goals.

In 1989, my wife Julienne (my college sweetheart) had been my most precious and faithful companion for 21 years. She continues to be so to this day and I fully expect so until 'death-do-us-part' (if not longer). My two great kids, Karl and Sarah, were then in high school, and now they are both married and out on their own. My life then and now continues to revolve and evolve about my family. My family life has been blessed, especially now with the recent addition of our first grandchild, Kade. He re-energizes my sense for life.

In 1989, I was in my sixth year as the '*Principal Electronic Engineer*' in the Corporate Research and Development Department at the W.H. Brady Company of Milwaukee, Wisconsin. It was a great job. I have always been an idea person, which can be either good or bad. It's good when new ideas are appreciated, and it's bad when the dominant concern is to not 'rock-the-boat' of the status quo. The W.H. Brady Company sought new ideas, so they appreciated my contributions and my three patents. Nevertheless, something was missing. I was not fulfilling my utmost potential for ideas.

In 1989, I was on an intellectual prowl, and I had read a number of books about physics. Two particular books contributed to an awakening in my thinking. The first book was Stephen Hawking's best-seller "*A BRIEF HISTORY OF TIME*". Dr. Hawking holds the 'Sir Isaac Newton's Chair' as the Lucasian Professor of Mathematics at Cambridge University, and this best-selling book catapulted Hawking into the public's eye. The second book was Clifford Will's "*WAS EINSTEIN RIGHT? Putting General Relativity to the Test*". Dr. Will is a Professor of Physics at Washington University in St. Louis, and he is a well-respected chronicler of the confirmations of General Relativity.

When I slid Professor Will's book over the counter to return it to our local library, the librarian quietly asked, *"Was he?"* I was already turning to walk away and I did not think the librarian was speaking to me. With no one else present, I then realized she must be speaking to me. She had asked, *"Was he?"* The first foolish thought that popped into my head was about a silly nursery rhyme: "*Fuzzy Wuzzy was a bear. Fuzzy Wuzzy lost his hair. Fuzzy Wuzzy wasn't fuzzy. Was he?*"

I then realized the librarian was speaking about the book, "*WAS EINSTEIN RIGHT?*" I froze in my half turned position and thought. With a couple moments lapsing in silence, the librarian must have thought I was some sort of snob. I then turned back, thrust my right forearm and forefinger upward, and boldly proclaimed to a somewhat startled librarian,

"NOT QUITE!
WAS EINSTEIN RIGHT? NOT QUITE!"

I did not try to explain my answer. I just felt filled with a sense of surprise at both my own boldness of expression and an awe of discovery. Disjointed ideas had seemed to flow together. I smiled at the librarian as the epiphany of the moment swelled within me into a new sense of awareness and purpose. As I turned to walk away, I knew my life was going to turn and follow a new path. I felt inspired by thoughts of *Einstein, Relativity,* and *Unification*.

The light bulb of thought that clicked on in my mind involved my developing perspective about the gravitational bending of light. Gravity bends light because it slows light. That was the key: Gravity slows light. The local measure of the speed of light is invariant because gravity also slows the local clocks and contracts the local measuring rods. Gravity does not curve spacetime into any 'extra dimension', but localized distortions of measuring devices do create a non-Euclidean geometry in three-dimensional space. In an extremely strong gravitational field, the bending of light can be so severe that light can follow a closed orbit. A photon (the particle of light) does not have a rest-mass but does have energy, but a photon in a small orbit would appear to be at rest from an outsider's perspective and thus possess an apparent rest-mass. Thinking in the sense that light is a wave of an electromagnetic field (within the continuum of space), I could thus envision how a field wave might form a rest-mass particle. I sensed this must somehow have been part of Einstein's thinking underlying his quest for a 'Unified Field Theory'.

At that time in 1989, my vision was quite vague about details, and I was not able to articulate it very well. These thoughts also seemed to be outside the mainstream of physics. Who would listen to me? I needed to learn more. Who would believe me? I realized the need to establish some credibility. For my mid-life redirection, I enrolled at the University of Wisconsin in Milwaukee (UWM) and subsequently earned both a Master's Degree and a Doctoral Degree in Physics. Now I am an Associate Professor of Physics at the Milwaukee School of Engineering (MSOE).

At MSOE, I introduced two elective courses, "*Relativity and Cosmology*" and "*Topics in Unification*". I have now taught these courses over the last four years. In addition to teaching the first course during our regular school year and during our last four summer sessions, I have also taught the second course three of our last four summers. Apparently, in the 102 year history of MSOE, my two courses are the only physics electives to run during any of MSOE's summer sessions. I believe this speaks to the wholesome curiosity of our current students who want to learn, as well as to the aura of mystery that surrounds these particular topics.

I have become so imbued with a positive focus for *Physical Unification*, that I have re-dedicated my life's work to pursue this goal and to write this book. The year 2005 is "*THE WORLD YEAR OF PHYSICS*" which celebrates the centennial of Albert Einstein's miracle year of 1905. The year 2005 is also the year I celebrate my 60th birthday. Not wanting my story to become lost due to my unexpected demise, I feel this is now the time to come forward. The year of 2005 is the year I choose to unveil what I envision regarding *Unified Physicality.*

As a member of the American Physical Society, I made two presentations at their April 2005 Meeting. The first session title was "*Perfecting $E=mc^2$*", and the second title was "*Debunking Instantaneous Non-Local Causality*". I offered too much material, far more than could be quickly assimilated. Moreover, the members were not prepared to receive such grand pronouncments. Overall, I was pleased with having established a precedence for my opening volley before a renowned body of physicists. The feedback from a few members has certainly been quite helpful.

To assure my precedence, I had copyrighted that April presentation along with a partial first draft manuscript of this book. At that time, I also trade marked the main title,

"WAS EINSTEIN RIGHT? NOT QUITE!" ™

Presentations at a conference and in many technical journals are limited because of standardized format constraints. However, a book provides a more open format that can enable fuller development of a larger topic. That is why I have chosen this book format as the primary vehicle for presenting this topic of *Unified Objective Physicality.*

Throughout all of this time over the past 16 years, my family and particularly my wife have been extremely patient and supportive. My job as an Electrical Engineer was certainly more rewarding financially. There have been moments when my wife Julienne asks, *"Are you sure this physics business is worth it?"* With a sassy look, I respond, *"For a mid-life crisis, don't you prefer I pursue physics rather than some other woman?"*

At some points, such as stimulated by my brother-in-law's death, I ask, *"What is the purpose of life?"* Since I began this journey in 1989, several relatives have passed away. Both my parents Paul and Conice, all my parents remaining siblings, my brother Robin, two beloved cousins Lloyd and Elaine, and a nephew John have all passed away. As the precious lives of loved ones go before us, it seems ever more important to not waste the life we have been granted.

I am a person of ideas. As a working engineer, I learned that good ideas are not worth anything unless you can put them into practice. My very nature compels me to seek new ideas and to put them into practice. This is the engineers' and physicists' version of *"Man does not live by bread alone."* Truths are goals onto themselves. Discovering the beauty of God's organization of nature can be an awesome experience. Those who believe they are blessed with the perception of new truths inherit an obligation to reveal those truths to others.

Personal Preface - Article 3

'Perfecting $E=mc^2$'

Throughout my adult life, I have involved myself with the ideas of energy. As a hobby, I have considered the issues of energy sourcing and energy efficiency. Like so many home owners, I have added extra insulation where I could get access. In the past, I have purchased and installed various gadgets for improving the efficiency of car engines and home furnaces. For many years, I have run a dual sourced water heater system that burns natural gas during the day and runs low cost off-peak electricity at night. I have drawn up numerous plans for solar and geothermal heating and cooling systems. I have studied the potential for using heat pump systems with both heating and cooling reservoirs. I have sketched innumerable home designs based on energy self sufficiency. I had formulated ideas for hybrid car drive systems years before they entered the market. For me, all this was a hobby, not a vocation, but an avocation.

Sometime in the near future, after the various editions of this book are complete, I would like to formally pursue the issues of improving energy sourcing and efficiencies. However, for now I need to get into my story about $E=mc^2$. In the 1980s, my avocation involving energy turned to more fundamental considerations. What are the limitations on energy efficiency? We cannot create energy, but we can manipulate it within limits. The 'Second Law of Thermodynamics' reveals the ultimate limit for extracting usable energy. In general, we are not nearly as efficient as we could be, and this inspired me to undertake study that was even more serious.

I then came upon an unusual story about a strange character, Immanuel Velikovsky, who is considered a 'scientific heretic'. His best selling book in 1950 "*WORLDS IN COLLISION*" sparked a scientific controversy (Venus and Mars nearly colliding with Earth during Biblical times). He predicted much - enough fantastically right to make him interesting and enough fantastically wrong to make him a heretic. I do not seek to justify or promote his work, though I was curious about his prediction of an enhanced role for electromagnetism within astrophysics. Otherwise, his efforts just struck me that reality may be much different than we have believed. In a series of communications with a somewhat doubting Einstein, Einstein kindly counseled Velikovsky. I found it somewhat curious that Velikovsky appears to be the last person with whom Einstein extended an ongoing communication on any technical matter before Einstein died in 1955.

Thinking about Einstein naturally leads one to think about his famous energy equation $E=mc^2$. In a convoluted way, Velikovsky thus led me back to energy. My new questioning of energy then led to questions that were more basic. What is energy? Where is energy actually stored? How does Einstein's famous equation $E=mc^2$ come to represent energy? In 1905 Einstein developed the energy equation $E=mc^2$ in which mass and energy are somehow equivalent. How does $E=mc^2$ relate to any practical issues about energy?

The nuclear fusion of Heavy Hydrogen seems such an ideal source for energy (abundant and non-polluting). Heavy Hydrogen possesses an extra particle (a neutron) in its nucleus, and possessing both a proton and a neutron, it is called Deuterium. We can readily extract Deuterium from water, which is composed of Hydrogen and Oxygen, H_2O. Though Deuterium appears in only small percentages, a gallon of sea water contains enough of it to generate fusion energy equivalent to the burning of 300 gallons of gasoline. Nuclear fusion can combine two Deuterium atoms into one Helium atom, which is an inert gas that does not pollute. The mass of the two Deuterium atoms is greater than the mass of the one Helium atom. Somehow, the missing mass is converted into energy via Einstein's energy equation $E=mc^2$. The sun uses this fusion of Deuterium to generate its energy, and we employ this fusion to explode Hydrogen bombs. However, we have not learned how to control this explosive 'Hot-Fusion'. We have not yet discovered how to slow the explosive 'Hot-Fusion' into a safe usable form called 'Cold-Fusion'.

Where is the energy stored when in the Deuterium? Supposedly, it is stored in the mass, so mass must be a repository of energy via the equation $E=mc^2$. The energy of fossil fuel is stored within its chemical bonds, but supposedly, the process of burning the fuel (combining it with Oxygen) also reduces the combined mass. Einstein revealed that motion also altered $E=mc^2$ and thereby manifested the classical kinetic energy of motion. What about gravity - where is the potential energy of gravity stored? Einstein did not appear to address this issue. In 1989, I came to realize that the potential energy of gravity is also stored within $E=mc^2$. Gravity alters $E=mc^2$. By altering the rest-energy of $E=mc^2$, gravity manifests gravitational potential energy. A raised object has increased rest energy and a lowered object has decreased rest energy.

What is the mathematical expression that describes gravity's alteration of $E=mc^2$? In developing that expression to include the potential energy of gravity, I feel that I am:

'Perfecting $E=mc^2$***'***

The energy equation $E=mc^2$ is still valid, but the details of its expansion become altered.

Einstein's Version: $\quad \boldsymbol{E = mc^2 = m_o c^2 \gamma}$

Gratke's Version: $\quad \boldsymbol{E = mc^2 = m_o c_o^2 \gamma_M^{1} \gamma_G^{-1}}$

Einstein introduced a special factor 'γ', (the lower case Greek letter gamma), to reflect the relative distorting effect of relative motion. I introduce two distinct distortion factors, 'γ_M' for absolute motion and 'γ_G' for gravity. (NOTE: In *Chapter 1*, we will introduce essential details to clarify these gamma factors.)

By 'Perfecting $E=mc^2$', I believe we position ourselves to perceive a fundamental insight for unifying Relativity and Quantum Mechanics. This long sought objective for a 'Theory of Quantum Relativity' now lies before us. The revelation is that both absolute motion and gravity alter the energy of quantum states that underlie the very formation of particles.

Personal Preface - Article 4

'Debunking Quantum Weirdness'

One of my earliest childhood memories outside of my family involves a tea party. I must have been 5 years old, and two little neighbor girls invited me to their private tea party. I loved milk and cookies, but if I had to drink some tea to get some cookies, I could manage to endure the tea. The girls had organized the site of their party in their cozy upstairs enclosed sun porch. As they ushered me in, I observed a small table all set with petite doll dishes. We took our seats around the table, and the older girl began to pour the tea. I saw NO tea pouring from the teapot's spout. With raised eyebrows and enlarged eyes, I looked at the two girls with clear surprise showing on my face. Ignoring my dumfounded expression and with no tea in their cups, the girls began to drink and urged me to drink as well. I lifted my cup and peered carefully inside down to its bottom. My cup definitely appeared to be empty, but was it actually empty? I carefully considered the question.

Now realize, at this time, I am only 5 years old. I am not considering the air in the cup to be something. Nor am I considering the 'photons of light', the 'aether of space', or the 'zero-point energy of a vacuum' to be something. Furthermore, I did not have the psychic powers to create that which was not there. Nevertheless, I wondered if these girls might have actually poured something into the cups. They were drinking whatever it was with clear satisfaction and no ill effect, so I tried it. After a moment or two of careful tasting and consideration, I concluded that the cups were indeed empty.

Either these girls were pretending or they had some sort of weird problem. If they were pretending, they must have practiced, because they were doing a very good job of it. On the other hand, if these girls were weird, I wondered about my safety. Maybe it was better for me that the cups were empty. In either case, when they passed around the 'supposed' plate of cookies, which was also empty, I realized I would be going home thirsty and hungry. These girls had obviously not yet learned that, *"The way to a man's heart is through his stomach"*. By mutual understanding, I never attended another of their parties. While I do realize these two girls were pretending, I cannot say the same for those who accept Modern Quantum Mechanics. They appear to reject *Objective Common Sense Physical Reality*.

We came to understand the legitimate role of Quantum Mechanics starting in 1899. We began to realize the internal structure of the atom. Electrons changing energy states produced unique spectral lines in their emitted light. Only the existence of certain energy states could produce the transitions corresponding to the colors observed for the emitted photons. The allowed energy states were called quantum states of the electrons, and the emitted photons were called the quanta of light.

However, many readers will be surprised to learn that Modern Quantum Mechanics has become extremely strange. Of course, physicists readily admit that Quantum Mechanics is 'Weird'. They admit they cannot explain why it should violate our common sense, but they assure us that it must be so. They have experiments which prove their propositions of 'Quantum Weirdness'. Consequently, Modern Physicists are unavoidably led to some very absurd points of view.

Carried to the extreme, the letter of the law of Modern Quantum Mechanics says that when no one is looking at an object it disappears. Thus, if no one is looking at the moon, the moon disappears. When no one looks at it, the moon dissolves into its 'quantum probability wave'. This non-physical 'quantum probability wave' spreads itself over the entire universe and also throughout an infinite number of 'parallel universes'. Somehow, all parts of this distributed 'quantum probability wave' are 'self-entangled' in an 'instantaneous' way. At the very instant that someone sneaks a peek, the moon 'instantaneously' reforms. 'Supposedly', all parts of the moon's 'probability wave' from the farthest reaches of an infinite number of 'parallel universes' will 'collapse instantaneously' to reform the moon. Thus, the one who sneaks a peek at the moon 'supposedly' causes the moon to reform into its familiar physical form of existence. This appears to be nothing less than psychokinetic 'thought controlled existence'.

When I confront non-physicists with this example of the dissolving moon, they usually do not believe that physics could be so foolish. I am sometimes falsely accused of making up this story about the moon, and therefore, I am falsely presumed to be some kind of charlatan, seeking to misrepresent Modern Physics. To remove such doubt, let me cite a reference. I first encountered this strange example of the moon dissolving in the early 1990s, when I was reading a back issue of "*PHYSICS TODAY*". In the April 1985 issue (pages 38 to 47), N. David Mermin presents an article titled in part, "*Is the moon there when nobody looks?*" "*PHYSICS TODAY*" is a highly respected Physics Journal, and Mermin is a well respected physicist. In the article, Mermin tries to explain the 'spooky' 'actions-at-a-distance' touted within Quantum Mechanics.

Einstein strongly objected to this 'spooky' model of Quantum Mechanics and said, *"God does not play dice with the universe."* He could not accept the 'anti-physical' qualities. Physical objects should possess the quality of 'physical existence' and 'physical persistence'. Physical objects should be able to exist independent of our observation and contemplation. Physical objects should not become involved in 'instantaneous nonlocal causality'. The dominant problem with Modern Physics is that the 'Standard Model of Quantum Mechanics' directly precludes *Objective Physicality* for all working models of the universe. Einstein thus objected to this preemption of his goal for a 'Unified Field Theory' which he envisioned to be based upon *Objective Physical Reality.*

On this point of objecting to 'Quantum Weirdness', I agree whole heartily with Einstein. Even at the age of five, I think I could have figured out the unlikelihood of the moon dissolving. At the age of five, I did figure out the empty teacup had no tea, and the empty cookie plate had no cookies. For a long time, I have thought there is something empty about 'Quantum Weirdness'. I have now discovered some key flaws within the formalities of Quantum Mechanics. I intend to show that the 'Standard Model of Quantum Mechanics' is based upon flawed theory and misinterpreted experiments. Hence, I will be fulfilling Einstein's goal of:

'Debunking Quantum Weirdness'

In 1935, Einstein (along with Podolsky and Rosen) proposed that a certain test might be performed to 'Debunk Quantum Weirdness'. All such proposed tests are now called EPR experiments (named after Einstein, Podolsky and Rosen). Physicists worldwide now recognize, that in 1981, Allan Aspect performed a definitive EPR experiment to 'Confirm Quantum Weirdness'. Thereafter, several other EPR experimenters have also 'Confirmed Quantum Weirdness'. Several years ago, I discovered what I believed to be a critical flaw in these EPR experiments, but I held my tongue. "*Bell's Inequality*", which is a critical factor in these EPR experiments, is simply WRONG! After careful reflection, I now believe, that with corrected interpretation, EPR experiments will actually 'Debunk Quantum Weirdness'. (NOTE: In *Chapter 1*, I will introduce critical details about Bell's Inequality and EPR experiments.)

In 'Perfecting $E=mc^2$' and 'Debunking Quantum Weirdness', I envision how we can combine Relativity and Quantum Mechanics and begin to develop a viable theory of *Physical Unification.*

Personal Preface - Article 5

'Cosmutonic Field Strings'

I remember that day in 1989 when I first learned of the death of my brother-in-law. When I arrived at home that day, my wife Julienne came out of the house to meet me. This was not her custom, so this in itself was an ill omen. She approached in a running walk. Her eyes were reddened and filled with tears. As I raised my arms to catch her in embrace, she blurted out, *"Ralph Evan is dead."* It was not a moment for further words. We comforted each other in the driveway with a long firm hug.

Ralph and I were not really close but certainly cordial and on good terms. Ralph was nearly 43 when he died, which seems much too young for dying. For the first time in my life, I felt the vulnerable mortality of my own existence. I had too many things I wanted to do, and now I felt I might never get around to them. For the funeral, we drove from Elm Grove, Wisconsin (a suburb of Milwaukee) down to Rock Island, Illinois (one of the Quad Cities on the Mississippi River). A long drive to a funeral provides much time for thought. On the quiet return trip, a sense of urgency paced my thoughts, which were wandering through a series of topics. Then I hit upon the an idea.

When it comes to getting an idea, I like the way P. G. Wodehouse expresses the process. In his book "*CARRY ON, JEEVES*" his fictional character Bertie Wooster exclaims: *"I pressed down the mental accelerator. The old lemon throbbed fiercely. I got an idea."*

I envisioned the idea of a photon in self-orbit to form an elementary particle, but I needed time to mentally digest this possibility. I subsequently learned that back in the 1950s, within a book entitled "*GEOMETRODYNAMICS*", John Archibald Wheeler had previously given the name of 'Geon' to a self-orbiting photon. However, the equations of General Relativity could not connect Wheeler's specific vision of a 'Geon' with specific elementary particles. I believe that Wheeler's basic idea of a 'Geon' was correct but he was working with incorrect equations. I believe we should employ altered equations for General Relativity, especially so at very short distances and also for very strong gravitational fields.

On page 91 of Einstein's book, "*THE MEANING OF RELATIVITY*", Einstein casually described his equations for gravity as being approximations. Unfortunately, everyone seems to ignore this very important qualification. They apply and extrapolate Einstein's equations as if they were exact. With a revised formulation of gravity, I believe the field formation of a particle takes on a totally new possibility. By 'Perfecting $E=mc^2$', I believe we can assert a new formulation of gravity that will enable us to discern the field formation of particles. By 'Debunking Quantum Weirdness', we can remove the main restraint against the idea of *Objective Physical Reality*, which I believe must underlie the *Physical Field* formation of particles.

Since ancient times, the idea of a 'continuum' forming the ultimate particles has vied with the idea of stand-alone particles dwelling within a 'void'. The 'continuum' was the medium of space and was called by many different names and was attributed to possess many various and conflicting properties. The 'void' was pure emptiness, and it possessed no properties because it was absolute nothingness. As we delve deeper into the structure of matter, our perception of the supposedly ultimate particles evolves to ever smaller sub-particles. For Democritus of Ancient Greece, the ultimate particles were 'atoms'. What we now call atoms are composed of electrons, protons, and neutrons. Then we were advised that protons and neutrons were built out of quarks. We also hear every sub-particle might be formed out of tiny little particles called Higgs Bosons. However, we never actually observe quarks directly, and the Higgs Boson is just an educated guess. More recently, we have been presented with speculations that the quarks or Higgs Bosons are tiny little 'Strings'.

Modern String Theory invokes the idea that tiny little Strings wiggle and jiggle to form the elementary particles. This is a nice idea. The Strings seem like they could be physical things that exist. Unfortunately, Strings are imbued with strange 'extra dimensions'. There have been several versions of 10 dimensional String Theory, and in 1995, Ed Witten appeared to unify all of them under the banner of his 11 dimensional 'M-Theory'. In addition to 'extra-dimensional Weirdness', Strings per se are the very antithesis of a 'continuum'. Einstein's 'Unified Field Theory' would have had the particles forming as disturbances in the continuum of the 'Total Field'. However, sub-particles (be they quarks, Higgs Bosons, or Strings) are viewed as objects in an empty 'void'. This is like comparing the negative of an exposed camera film versus the developed picture. Strings are the negative perception, and the 'Total Field' is the positive *Physical Reality.*

We need to be careful and to clarify terms. I label the medium of space as being the *Cosmic Continuum*. Being like a solid, physical, elastic medium of space; the *Cosmic Continuum* can become distorted to form a *Physical Field* (the physical version of an abstract field). This *Physical Field* should require a mathematical form comparable to Einstein's vision for the 'Total Field'. Whirlwind waves within the *Physical Field* can form *Solitons*, which are the elementary particles. A *Soliton* is a self-perpetuating wave that can form in a non-linear medium. These *Solitons* can move through and interact with the *Cosmic Continuum* and with other *Solitons*. By virtue of these interactions, these *Solitons* can thus '*Mutate'*. I label these entities *Cosmutons*, an acronym representing "*COSmic Continuum MUtable SoliTONs*". (The accent is on the first syllable, ***Cos'-mu-ton***).

Dynamic distortions of the *Cosmic Continuum* can wiggle, jiggle, twist, and rotate. At some point the distortion forms into a *Cosmutonic* vortex, somewhat like a miniature tornado or hurricane. A *Cosmuton* thus involves distortions in three-dimensional space, and we can identify three regions: an inner region, an outer region, and a boundary region between the inner and outer regions. The boundary region of a *Cosmuton* takes on the mathematical qualities of a String. Hence, I label the active boundary region of an elementary particle as being a:

'Cosmutonic Field String'

In the cover artwork, the artist Shawn Tuft seeks to depict a mutating rest-mass *Cosmutonic Field String* (but not to scale). This could be an electron emitting or absorbing a photon depending upon which way you view the reversible process to be running. This closed-loop helical String looks like a French donut with several twisting lobes. Each lobe represents one wavelength, and the closed loop must consist of an integer number of wavelengths to maintain harmonic resonance. The total number of lobes constitutes the number of the current quantum state. More lobes means more energy. By adding a lobe, we increase the harmonic quantum state. By taking an extra lobe away, we decrease the harmonic quantum state. The cover artwork also depicts a non-rest-mass *Cosmutonic Field String*. This could be a photon (the particle of light). It looks like an open loop String with a twist. When an electron mutates to drop one quantum state, the electron breaks off one of its extra twisting lobes to create an escaping photon. Conversely, if a photon becomes entangled with an electron, the electron assimilates the photon by adding an extra twisting lobe to increase the electron's quantum state.

David Bergman (cofounder of "*COMMON SENSE SCIENCE*") has been paralleling my research for *Cosmutonic Field Strings*. In 1990, Bergman suggested a spinning electric-charge ring model for an electron and proton. His group now calls this kind of particle a 'Helicon'. For Bergman, electric charge is supposedly the elemental substance underlying all natural phenomena. I disagree because I envision the *Cosmic Continuum* as underlying all natural phenomena. Nevertheless, I like Bergman's ring model because it outwardly resembles my model of a *Cosmutonic Field String*. While the internal underlying physicalities of the two models are completely different, the exterior qualities of the two models possess some similarities. Consequently, they lead to some similar descriptions for the buildup of matter.

All elementary particles are formed out of the same substance. They are formed as disturbances of the *Cosmic Continuum*. As mentioned before, all elementary particles are created as "*COSmic Continuum MUtable SoliTONs*" now to be called *Cosmutons*. The easiest feature of a *Cosmuton* to visualize is its self-interfacing boundary, now to be called a '*Cosmutonic Field String*'. Previously, 'Strings' were imagined to be entities in and of themselves, but '*Field Strings*' are states of distortion of the *Cosmic Continuum*, the physical medium of space.

Personal Preface - Article 6

'The Cosmic Continuum'

In 1959, I was a freshman at Bay View High School in Milwaukee Wisconsin, and the Lewellyn Public Library was conveniently located across the street. I had checked out a book about Relativity. My mind is one that visualizes things, and in this case I remember the color, texture, size and shape of the book to this day. It was a gray cloth hard covered book about 6x9 inches, and it was thin. Including the cover, the thickness of the book was about half an inch. I was a slow reader, so fewer pages seemed attractive. I must have checked out this book on Relativity for some book report that was required. Why else would a physically active 14-year old check out a library book? For some book reports we had the option of giving an oral report, and I would always do the oral reports because that was less work for me. I do not remember anything about the book report, but it must have made an impression on at least two of my fellow students.

A few years ago, I was paging through my high school year book from that year of 1959. It was our custom back then to write little notes of remembrance to one another on the leading or trailing blank pages. The girl who sat next to me in homeroom had written "*To Einstein II*". In the early 1990s I met by chance another high school classmate. He informed me he had become an Electrical Engineer. I countered that I also had become an Electrical Engineer but was now studying for my Doctorate in Physics. He expressed his surprise, not at me seeking a Doctorate in Physics, but that I had studied Electrical Engineering. He thought I always intended to be Physicist and did not think I would have gone into Engineering.

My brother Robin (five years older) passed away ten years ago, which only added to my own sense of mortality. As teenagers, we had worked or several home projects including electrical work and even hired out to do electrical work for others. When Robin had signed up to study Electrical Engineering, I just followed his lead. I had never given any thought whatsoever to physics. Physics was just some fun stuff to think about. Who would hire a person to do physics to fix up their house?

Now, let us get back to that book about Relativity (more specifically, Special Relativity). Having read it, I formed the opinion that space distorted objects in absolute motion to create the invariance of measurement, but due to that invariance, absolute motion could not be detected. This was the view of Henri Lorentz from the 1890s. Lorentz thus explained the null result of the Michelson-Morley Experiment of 1887, which attempted but failed to measure absolute motion. Modern Physics considers this Lorentzian view to be wrong. Nevertheless, since 1959, I have continued to hold and support the view articulated by Lorentz. In the case of Special Relativity, the disagreement seems to be a case of semantics. However, in rejecting the Lorentzian view, one also rejects a physical medium of space, and this is the great mistake of Special Relativity.

Had I gone into physics as a young man, I believe the academics would have crushed my good sense and my zeal. By waiting until I was older to formally study physics, I became more secure in myself and my understandings. How does light know to go at only one speed? If light is composed of particles called photons, why do the photons go at only one speed? If light were a wave, it would make sense that it goes at the speed of a wave through the medium of space. That then becomes the number one question. What is the medium of space?

Once we understand that elementary particles are dynamic formations within a physical continuum, we then transform our whole perception of natural phenomena. A solid medium of space had always been a problematic proposition because it would appear to prevent the free motion of objects. How could planets move through a solid medium to orbit the sun? However, now we can perceive that objects such as planets are composed of little waves within the solid physical continuum and can thus move with no friction whatsoever (like particles in a void). In addition, the continuum itself can now communicate forces without the need of force mediating particles as currently envisioned by today's physics.

To avoid any confusion with the previous incomplete descriptions, I needed a new distinctive expression to designate the physical medium of space. When first trying to decide what to call this 'stuff of the cosmos', my wife Julienne (as always) provided valued constructive feedback in selecting the name '*Cosmic Continuum*'.

The various incomplete and incompatible 'aethers of space' have created an extremely negative impression for the very idea of a medium of space. Today's physicists simply reject all 'aethers'. They then talk about 'the zero point energy of the vacuum', which is a deceptive way of describing something about nothing. 'The zero point energy' is something, 'the vacuum' is nothing, and together they mean 'aether' without having to say the forbidden word 'aether'. In any case, I disagree with all of these past descriptions for fields and the medium space because they are 'NOT QUITE' right. I envision the *Cosmic Continuum* to be like a solid elastic compressible physical medium, and it fills and forms our single universe of three spacial dimensions.

In 1989, I discovered my initial vision of a *Cosmuton*, though I did not come up with that name until several years later. Regardless of its name, I was thinking in terms of rest-mass particles forming somewhat like photons in self-orbit. This raised six interesting questions, which I have ever since sought to resolve.

One relates to the 'variable' speed of light, another relates to the 'true nature' of elementary forces, a third relates the 'force reaction' of the medium of space, a fourth to 'dark matter', a fifth to 'blackholes', and a sixth to 'quasars'.

- For a wave to travel in an arc or curved path, the wave front must be traveling progressively slower toward the side in which the wave direction bends. Progressive wave slowing is the essential factor to form a curved path of propagation. Einstein and Sir Arthur Eddington both thought that gravity progressively slows incoming light, but almost all of today's physicists seem to ignore this *Physical Reality*. The question then becomes, *"What are the factors that might cause a wave to travel in a loop?"* The answer is, *"The variation of compression density of the medium induces a wave to travel in a loop."* The variation of the speed of light within the *Cosmic Continuum* has been the key for me to unravel the mystery of particle formation.
- Both Newton and Einstein thought in terms of forces communicated through a medium. For Newton, forces were induced by variation of density of a 'gaseous aether'. For Einstein, forces were induced by variation in the 'space-time continuum' of the 'Total Field'. I believe static and dynamic variations of compression within the *Cosmic Continuum* induce the elementary forces. I believe that elementary particles form as unique waves within the *Cosmic Continuum.* Just as gravity bends the path of light, distortions of the *Cosmic Continuum* bend the paths of a particle's internal waves to manifest the effects of all the truly elementary forces.
- If the medium of space is the communicator of forces, then Newton's third law - the Law of Reaction - must apply. If the medium pushes or pulls on an object, then the object must push ar pull back upon the medium. The medium must then be altered in some way in response to the reaction. This reaction should reduce the original force, but how much? Does this ever reveal itself?

- The anomalous orbital rates of galactic stars have been attributed to the existence of 'dark matter'. However, does this dark matter really exist? In the case of galaxies, the orbiting stars constitute the dominant mass of a galaxy to create its gravitational pull. However, by virtue of orbiting, the masses impart a reaction force pushing outward on the medium. This *Gravio-Reaction Force* thus alters the nominal gravitational field, and this alters the orbital rates of the galactic stars. That dark matter, which was conjured up solely to account for the anomalous stellar orbits, does NOT exist!
- If particles were like photons in self-orbit, how could particles fall below the photon orbit of a 'blackhole'? Supposedly, the 'horizon' of a 'blackhole' forms below the photon orbit of a 'blackhole'. The very idea of a 'blackhole' thus seemed dubious. Einstein had said that he thought 'blackholes' were un-physical. My formulation of $E=mc^2$ also suggests that 'blackholes' do NOT form. If 'blackholes' do NOT exist what takes their place. I call them *Greyholes* because they emit light, but very dimly, and the light is greatly redshifted.
- A few years ago, I was delighted to realize some astrophysical objects looked like my predicted *Greyholes*. Quasars are just *Greyholes*, which have been mischaracterized. Quasars are falsely depicted as being the most distant objects in the known universe. We shall later see how this discovery vindicates the much-maligned Halton Arp. Arp is an astronomer who has persistently claimed that quasars are close to galaxies in direct conflict with the standard interpretations of Modern Physics.

To ignore the variation of the speed of light has served to obstruct the proper discernment for a physical medium of space. That was my greatest objection to Clifford Will's book "*WAS EINSTEIN RIGHT?*", because it seemed to let physicists play semantic language games that denied the variation of the speed of light. To deny a physical medium of space has served to obstruct the proper discernment of the most fundamental *Physical Reality*. The *Cosmic Continuum* is the physical medium of space and is the ultimate substance of the universe. Quite simply, the *Cosmic Continuum* underlies all natural phenomena.

Personal Preface - Article 7

'The Holy Grail of Unified Physicality'

Albert Einstein died in 1955, but at the time, the event of Einstein's passing escaped my notice. I was just ten years old, and I was enthralled with the opening of Disneyland. I did not get to go to Disneyland as a youngster, but we all experienced it through the television program "*The Wonderful World of Disney*". Watching the televised development of Disneyland created a strong sense of mankind's ability to design and build a better world - if only we would focus our attention on the task.

As a youngster in the early fifties, I liked to see things organized. Once I thought my mother's kitchen cabinets were cluttered, and I wanted to surprise her by re-organizing them. In her absence, I emptied all the cabinets, cleaned the shelves, and neatly placed items back where I thought they should go. My expectation of my mother's joyous surprise was somewhat diminished when she saw the kitchen table filled with so many items left over, which did not fit into my re-organization. We obviously needed more kitchen cabinets. (A few years later we did remodel to add extra cabinets.)

On another occasion, my father was also the recipient of my good intentions to re-organize. My father was a librarian and had acquired many books for his own library, a total of about 10,000 books. One day, while he was at work, I decided to re-organize some of his library. The criteria I applied was color. All the red books here, all the blue books there, and the same for all the other colors. Within color groups the different shadings of colors were sub-grouped. It had a wonderful rainbow effect. Having six kids in the house, my father had developed a good sense of patience, but I believe that day must have seriously tested the limits of that patience. Surprisingly, he did not get outwardly upset. I suspect he was hoping that a process of osmosis might occur, such that in handling the books, the ideas they contained might begin to seep into me.

To organize or to re-organize things can be a wonderful experience, or not so, depending on your perspective of the given plan of organization. When it comes to understanding natural phenomena, we do not re-organize and design, but we try to perceive the underlying organization which we presume must exist. We observe, we analyze, and we guess what the organization might be. We use our organizational skills to project what the existing hidden organization might be. We then test our guesses. We discuss, we argue, and we re-assess. Then we guess again. If we are dedicated, disciplined, patient, persistent, thoughtful, thorough, and competent; then we are doing the work of science and make progress.

The heart and mind ever yearns to know the truth about reality. What is real? What is not real? How can we be confident about what we believe about reality? From antiquity onward, some of the greatest minds of recorded history have struggled to perceive an underlying unification for all natural phenomena. Since no one heretofore has realized this cherished dream, we can be quite sure that this is not a trivial pursuit. At times, '*Ultimate Unification*' certainly seemed to be an impossible dream, but the human spirit can be resilient and optimistic.

The greatest physicists in the world today look forward with positive expectations for a future '*Theory of Ultimate Unification*'. The Physics Nobel Laureate, Steven Weinberg, has written a book entitled "*DREAMS OF A FINAL THEORY*". The world-renowned physicist, Stephen Hawking, has written a book entitled "*A THEORY OF EVERYTHING*". They are trying to carry the torch that Albert Einstein ignited. Though unsuccessful, Einstein ambitiously sought to develop a 'Unified Field Theory'. I describe this search for *Ultimate Unification* as a quest for 'The Holy Grail of *Unified Physicality*'.

The existence of life itself and its origin are the two greatest mysteries of science, but the story of life is outside the scope of this work. The existence of non-living physicalities and their origins are the third and fourth greatest mysteries of science. However, the contemplation of origins constitutes a never-ending topic of intriguing speculation. Rather than contemplating guesses about long ago primordial events, this work concentrates upon what exists now. Physics is the study of the existence and characteristics of non-living physicalities. The underlying unified nature of these physicalities in today's universe is the primary subject of this work.

The quest for discerning the underpinnings of nature reaches back to ancient times. In our modern era, we sometimes forget the wisdom of the ancients and those that followed. We smugly think how advanced we have become, when in fact history reveals we should be somewhat humble. In ancient Greece, Democritus and Aristotle respectively touted contrasting views of particles and fields. In the 1600s, Newton unified terrestrial and celestial mechanics, and (as Einstein himself reported) Newton touted a wave-particle-duality theory of light. In the 1800s, James Clerk Maxwell unified electricity, magnetism, light, and thermal radiation within his Theory of Electromagnetism (a wave theory of light).

In the 1900s, Einstein explained the photoelectric effect via photons (the quanta of light) to re-introduce a wave-particle-duality theory of light. Einstein then searched in vain to merge Newton's mechanics and Maxwell's electromagnetism into a complete theory of unification. The idea of unification seemed so close and tantalizing. That is why Steven Weinberg "*DREAMS OF A FINAL THEORY*", and Stephen Hawking anticipates "*A THEORY OF EVERYTHING*".

It seems fair to say that the goal of unification can be called 'The Holy Grail of *Unified Physicality*'. It stands as the greatest mystery in all of the history of Physics. Regarding non-living physicalities, some basic questions are as follows. *"What exists? How does it exist? Is there one thing that exists, or are there many things that exist?"* Einstein envisioned a 'way of thinking' about physicality via his hopefully unifying abstract 'Total Field' (a single entity). That view has not only drifted out of vogue, but it has been driven out of academia.

Einstein sought a 'Unified Field Theory' governed by an *Objective Physical Reality*. Modern Physicists have been pursuing the idea of particles (such as quarks and Strings) governed by 'Extreme Quantum Weirdness'. If this were just a contrast of a single continuum versus many particles, but both within an *Objective Physical Reality*, then we could have a good-hearted debate. Both approaches would then exude a decent sense of *Physical Reality.* However, the real issue is between *Objective Physical Reality* and 'Extreme Quantum Weirdness' and this conflict cannot be treated with caviler indifference. While I agree with Einstein in promoting a continuum, the greater issue is that I agree with Einstein in promoting *Objective Physical Reality.*

It will surprise many that in my seeking of an *Objective Common Sense Physical Reality*, I have discovered:

'The Holy-Grail of Unified Physicality'.

This pioneering work goes directly to the essence of Einstein's ideas. It overcomes some absurd aspects of Modern Physics just as Einstein had sought. It recasts the whole of physics into a context of *Common Sense Physicality* that underlies all natural phenomena. This physicality involves nonlinear dynamics of the *Cosmic Continuum* governed by *Localized Causality* in today's single universe of just three spacial dimensions. I believe my '*Unified Physical Field Theory'* begins to realize Einstein's dream for a 'Unified Field Theory'. As such, I believe we are now poised to enter into a new era - '*The Era of Unified Physicality'*.

Personal Preface - Article 8

An Original and Historic Scientific Treatise

For 16 years, I have been dedicated and focused upon developing my ideas and my credentials for announcing an *Objective Common Sense Physical Reality.* Now, herein, I propose to outline nothing less than the discovery of 'The Holy Grail of *Unified Physicality*'. I believe, that in seeking *Unified Physicality*, I am pursuing the development of an original and historic scientific treatise.

I contend Einstein virtually 'WAS RIGHT' about the observations related to his Relativity, but he was 'NOT QUITE' clear about explaining its underlying physicalities. Here are what I perceive to be the most prominent flaws in the legacy of Einstein's Relativity:

- Einstein's Special Relativity has confounded our common sense. How can two clocks both run slower than each other? Riddles such as this unnecessarily tease our sense of good logic and suggest an improper aura of mysticism.
- Einstein's Special Relativity appears to preclude the measurement of absolute motion, but this is misinterpreted to dismiss the idea of a physical medium of space. It can be argued that relativistic observations indirectly affirm the existence of a physical medium.
- Einstein's General Relativity is misinterpreted to ignore that gravity slows incoming light. Gravity does slow incoming light, and (as noted below) this portends a new understanding of curved spacetime and $E=mc^2$, as well as the renunciation of 'blackholes'.
- Einstein's General Relativity is typically misinterpreted to depict space as curving into an 'extra dimension' (the 'wormhole fifth dimension'). Space does NOT curve into any 'extra dimension'. Gravity is formed as a *Compression Density Gradient* of the *Cosmic Continuum*. Objects distorted in three-dimensional space form a non-Euclidean local geometry.
- Einstein's General Relativity failed to properly describe the effects of gravity upon the energy of photons and the energy of ordinary matter. Gravity does not alter the $E=hf$ energy of photons in transit, but gravity does alter the $E=mc^2$ rest-energy of ordinary matter.
- Einstein's General Relativity is seriously misinterpreted to depict formation of 'blackholes'. Einstein warned that his formulas were only approximate and that 'blackholes'

were un-physical. Very dense objects are henceforth to be called *'Greyholes'*.

- Einstein's Relativity failed to include its effects upon the quanta underlying Quantum Mechanics. Relating the magnitude of quanta to absolute motion and gravity now leads to a unification of Relativity and Quantum Mechanics.

I contend Einstein 'WAS RIGHT' but incomplete about physicalities underlying Quantum Mechanics, so he was 'NOT QUITE' able to formulate a winning argument. Einstein's efforts to overcome the physical absurdities of Quantum Mechanics are legendary, but since he appeared to lose his arguments, he and his cause suffered rebuke. I believe the following physical absurdities of Quantum Mechanics and String Theory are simply unbecoming of science and they are WRONG:

- The 'Heisenberg Uncertainty Principle' misinterprets 'quantum probability waves' to deny 'physical persistence' and thus deny 'physical existence'.
- 'Quantum probability waves' are misinterpreted so that objects supposedly persist in a dissolved and distributed state of 'non-existence' until observed.
- These dissolved and distributed states of 'non-existence' are foolishly thought to dwell within an infinite number of 'parallel universes'.
- Objects distributed in these 'parallel universes' are imagined to be 'mystically connected' via some strange 'instantaneous non-local entanglement'.
- Unbelievably, 'instantaneous non-local entanglement' enables 'instantaneous non-local causality' throughout all of the make-believe 'parallel universes'.
- When we attempt to observe an object, the dissolved and distributed state of 'non-existence' supposedly collapses 'instantaneously' into the object we see.
- Misinterpreted 'EPR experiments' claiming to prove 'instantaneous non-local causality' are the greatest self-deception in the history of science.

- The 'extra-dimensionality' of String Theory is a transgression of mathematical semantics that ignores and denies realistic physicalities.

I contend Einstein essentially 'WAS RIGHT' in his vision for unification, but he was 'NOT QUITE' able to overcome the flaws and the accompanying 'mathematical obfuscation' of Modern Physics. Once we overcome the flawed aspects of Relativity, the strange tenets of Quantum Mechanics, and the semantic miscues of String Theory, we can then reflect anew. I believe we can reset the foundation stones of physics. Contrary to current common belief, we can establish a new vision of *Unified Objective Physicality*.

My vision of *Unified Physicality* is that our universe dwells within three spacial dimensions and is composed of a single physical substance, which I call the *Cosmic Continuum*. Distortions of the *Cosmic Continuum* constitute a *Physical Field*. These *Physical Field Distortions* can form self-supporting rotating wave structures to form what we have previously identified as the elementary particles. These whirlwind wave formations of particles can be considered to be *Field Strings*, but I prefer to call them *Cosmutons*.

I have coined the term *Cosmuton* as an acronym. Let me explain this acronym for a third time because it is so important. Fully understanding the acronym, greatly assists in understanding the concept of particle formation. *Cosmuton* represents "*COSmic Continuum MUtable SoliTON*". (The accent is on the first syllable, ***Cos'-mu-ton***). In general, a 'soliton' is a self-sustaining wave within a non-linear medium. A *Cosmuton* is a 'soliton' forming within the *Cosmic Continuum*. A *Cosmuton* is quantized because it is a wave in self-resonance in a non-linear medium, and such non-linear resonance requires quantized amounts of energy. A *Cosmuton* is 'mutable' in two ways. Absolute motion and gravity can both smoothly shift the amount of a *Cosmuton's* resonant energy. Also, in qualified ways, a *Cosmuton* can absorb or emit another *Cosmuton* to incrementally shift its resonant energy.

The primary wave resonance takes place within a *Cosmuton's* inner *Core Region*, and that is what we would nominally identify as a particle. However, a secondary action takes place in an outer *Aura Region*, so that a *Cosmuton* manifests an extended 'wave-particle-duality'. *Cosmutonic Auras* utilize the physical medium of the *Cosmic Continuum* to form the various force fields without any mediating force particles. In my *Unified Physical Field Theory*, I thus seek to unify Relativity, Quantum Mechanics, String Theory, Classical Mechanics, and Classical Electrodynamics.

I believe my vision of *Unified Physicality* will transform our perception of the most basic structures in our universe. On the microscopic scale, we will significantly reform our understanding of atoms and elementary particles. On the macroscopic scale, we will develop surprising new understandings about galaxies, quasars, and dark matter. The development of the *Unified Physical Field Theory* is nothing less than the discovery of 'The Holy Grail of *Unified Physicality*'.

Personal Preface Post Script: As you delve into this proposed discovery of 'The Holy Grail of *Unified Physicality*', read as though you are reading a mystery novel. Put yourself into the character of Sherlock Holmes, and take up the challenge of trying to see how the pieces of this puzzle fit together. Open your mind. Witness that *Common Sense Physicality* can indeed underlie all natural phenomena. Recognize the truly great revelation that *Objective Common Sense Physicality* is being rightfully restored as the basis of science.

Prologue

Building Upon Einstein's Legacy

"Was Einstein Right? Not Quite!"™
But Einstein was Oh-So-Close.

Albert Einstein spent most of his adult life seeking to develop his dream of a 'Unified Field Theory'. Einstein envisioned that his 'Total Field' formed all the forces of nature and also formed all the elementary particles of nature. I agree with this vision of *Physical Reality*, and I seek to build upon Einstein's legendary vision.

Prologue - Article 1

Centennial of Einstein's Miracle Year

The year 2005, "*THE WORLD YEAR OF PHYSICS*", is the centennial of Albert Einstein's miracle year of 1905 and the fiftieth anniversary of his death in 1955. It is most proper to now honor Einstein for his great ingenuity and wisdom. However, let us not fall into the mistakes of the past that bestow honors on him based upon flawed perceptions. Einstein's legacy of Relativity is slightly overrated, his questioning of Quantum Mechanics is greatly underrated, and his vision for unification is generally misunderstood. Let us now bestow a special honor upon Einstein that, heretofore, has been denied him.

Let us honor Einstein for his vision of unification. What better way to so honor him than by fulfilling his cherished dream of a 'Unified Field Theory'. Herein, I thus propose to introduce my own vision of '*Unified Physicality*'. However, rather than seeking unification via an 'Abstract Field', as did Einstein, I seek unification via a *'Physical Field'*. I consider my vision to be a scientific revelation, and I call it the *'Unified Physical Field Theory'*.

Physicists have denounced Albert Einstein in one respect, because he advocated *Objective Physicality* for Quantum Mechanics and unification. I believe Einstein 'WAS RIGHT' about this hidden physicality, but he was 'NOT QUITE' able to convince his contemporaries. A critical factor that hindered his quest for unification was that he was 'NOT QUITE' right about Relativity. Somehow, Einstein let himself be trapped by the mathematical elegance of his Relativity theory. He thereby misguided us and led us away from the underlying physicality for Relativity and force fields, and thus he led us away from the *Objective Physical Reality* underlying Quantum Mechanics and unification!

With all due respect to Einstein, his ideas were simply not complete, and this has left Einstein's ideas in a state of some confusion. Some of his best ideas were wrongly rejected. Other great ideas of Einstein, which are essentially correct and are accepted, lack a strong sense of an underlying physicality, and thus exude an improper aura of mysticism. More seriously, some of his good ideas are wrongfully applied to conjure up utter fallacies. I propose to now offer a new perspective of *Objective Physicality* that clarifies, justifies, corrects, and completes Einstein's work. I thus seek to demystify Einstein's Relativity, to vindicate his vision of quantum physicality, and to achieve his splendid dream of unification.

Prologue - Article 2

A New Era of Unified Physicality

Since ancient times, we have also had to contend with the conflicting visions for underlying physicality based upon the ideas of particles versus the ideas of a continuum. In Ancient Greece, Democritus proposed atoms in a void, while Aristotle proposed the aether continuum of space. Most Modern Physicists think in terms of various ultimate sub-particles such as quarks and Strings, which in vast numbers can create the appearance of a continuous field. In direct contrast, Einstein thought in terms of a continuous field that could form disturbances, which create the appearance of particles. I agree with Einstein, and seek to present the *Cosmic Continuum* as the basis of *Ultimate Unification*.

The task of *Ultimate Unification* has been extremely difficult, in part because the basic theories to today's physics do not fit together to create a cohesive whole. Today's physics is built upon a foundation with two fundamental theories, 'Relativity' and 'Quantum Mechanics'. Relativity involves continuous dynamics and tends to describe large-scale effects, and Quantum Mechanics involves discrete dynamics and tends to describe small-scale effects. These two fundamental theories as they are currently constituted are not compatible. Modern Physicists have long sought to reconcile this conflict and to develop a theory of *Quantum Relativity*. I believe my *Unified Physical Field Theory* addresses this goal of a unified *Quantum Relativity*.

The great difficulty facing any attempt to develop a vision of *Unified Physicality* is to simply recognize the actual flaws in our current conflicting perceptions. Somehow we have misunderstood and we have miss-characterized some aspects of the most basic tenets of physics. I believe the flaws lie before us in open view, but we have failed to recognize them. Some very simple and yet very subtle truths await our discernment, and I propose to point them out. When the story is finally understood to be so basic and so reasonable, many will correctly wonder,

How could we have allowed ourselves to become so confused?

I believe the reason we have not heretofore developed a viable vision of unification is that we are too committed to partial truths that lead us away from the whole truth. It is like a magician's slight of hand that misdirects our attention. The enormity of the current status-quo rationale for 'anti-physical dogmas' constitutes a monumental barrier that obstructs any significant change for the better. I believe physicists will overcome initial resistance to eventually understand and appreciate my vision of *Common Sense Physicality*. The task of developing a *Theory of Ultimate Unification* must confront three monumental problems.

- The first problem is to overcome the current conflicts of Modern Physics, and this requires exposure of the fallacies within Modern Physics.
- The second problem is to actually develop a self-consistent physical theory of unification that is compatible with known observations.
- The third problem is to overcome the naysayers who will not consider any variation to currently established abstract mantras.

Others have previously suggested various forms for some aspects of my revelations for *Common Sense Physicality*. However, those previous suggestions have been incomplete, so they are conveniently ignored and/or characterized as invalid. Since most common sense pronouncements about physics are indeed wrong, it is almost unavoidable that the few valid ideas are overlooked. Prior common sense pronouncements thus appear to have had no impact on today's 'anti-physical' thinking. Today's physicists have developed an immunity to any proposed foundational ideas, which appear to be based upon *Common Sense Physicality*.

When anyone conveniently ignores a reasonable common sense proposition and argues instead for an apparently contrived 'anti-physical' proposition, we have to wonder why. Given a clear choice, why would anyone not accept an explanation based on a common sense perception? The answer lies in the habitual ways of thinking that tend to form a closed mind. We hear but do not listen; we look but do not see; and we contemplate but do not understand. A few incorrect facts are then 'unintentionally manufactured' to support the extreme visions touting 'anti-physicalities'. Blinded to the viability of *Objective Physicality*, the opposing logic thus employed to support 'anti-physical dogmas' becomes tainted because it ignores reasonable alternatives. That kind of thinking is called sophistry, which can be thought of as the abusive negative use of philosophical argumentation.

Arguing with great collective skill, but like sophists with little wisdom, today's physicists blindly promote an unacceptable 'sophistry of anti-physicality'. I believe this well entrenched 'sophistry of anti-physicality' is wrong and is simply unworthy of science! However, this 'way of thinking' does have its justifications. We do have grand mathematical formulations and experiments that supposedly 'prove' the mysterious pronouncements of today's physics. I contend we have misunderstood those mathematical formulations and we have misinterpreted those experiments. We are led by flawed mathematical formulations to accept 'non-physical' and 'anti-physical' abstractions, and thereby we fail to see the physicality that actually exists.

Much of today's physics reveals valid truths, but some aspects of its well-entrenched basic tenets are unbecoming of science. I contend those aspects of Modern Physics which constitute 'anti-physical dogmas' are offensive to the very spirit of physicality and are simply unacceptable. Some other errors in today's physics are not offensive to the sensibility of physicality, but they are errors nonetheless. The most egregious errors will be exposed in this *Monograph* and in greater detail within the *Full-Book Edition*. The bulk of what constitutes Modern Physics will remain unaltered, but that which is altered is so basic that it will appear as if we are entering into a new era of physics. We could thus say this work proposes to transition us into a new era - '*The Era of Unified Physicality*'.

Prologue - Article 3

Einstein's Guiding Wisdom and Warning

Albert Einstein was and continues to be a 'Great Spirit'. The cover artwork drawn by Shawn Tuft seeks to depict Einstein's ever-present 'Great Spirit'. As a student, Einstein seemed somewhat distracted and preoccupied, and he did not focus well on his assigned material. He was ranked lower than most of his fellow students, so he was not even recommended for an academic position after completing the course work for his doctorate degree. In fact, special appeals had to be made on his behalf to secure for him the position of patent clerk. As a patent clerk, he quickly and competently completed his tasks. With time left over, Einstein could secretly work on his grand ideas while sitting at his desk in the patent office. Einstein later quipped that the top right drawer of his desk at the patent office was his "*Department of Theoretical Physics*".

In his miracle year of 1905, the brilliance of this patent clerk blossomed. In that year, Einstein published six technical papers, four of which have become classical landmarks in the advancement of science. The topics of the four famous papers were:

- Brownian Motion (the thermal agitation of atoms and molecules induced by collisions).
- The Photoelectric Effect (with E=hf for the energy of a photon, the particle of light).
- Special Relativity (the relativity of motion with constancy for the measured speed of light).
- Mass-Energy Equivalence (with $E=mc^2$ for the energy of matter to complete Special Relativity).

In 1916, Einstein extended his Special Theory of Relativity into his greatest achievement, the General Theory of Relativity (the relativity of gravity and curved spacetime).

Anticipating unpleasant negative responses when I began my research in earnest sixteen years ago in 1989, I posted two notes on my kitchen cabinets quoting Albert Einstein. Discoloring at the edges with age, the still posted quotes reveal Einstein's wisdom.

Einstein's first quote reads:

"For an idea that does not at first seem insane, there is no hope."

How could any one person dare to think they alone could unravel the mystery of Einstein's dream for a 'Unified Field Theory'? It does indeed 'seem insane', especially when coming from a previously unknown researcher such as myself. However, without a grand vision, 'there is no hope'. I found encouragement and comfort in Einstein's admonition offered within this first quote.

Einstein's second quote reads:

"Great spirits have always encountered violent opposition from mediocre minds."

A few days after posting the second quote on my kitchen cabinets, it finally dawned on me that I needed to offer some clarifications. A rather subdued silence from my wife following the message posting indicated that misunderstandings can occur at many levels. I was thinking to apply the term *'mediocre minds'* to mathematically minded Modern Physicists who zealously promote the 'sophistry of anti-physicality' - NOT to physically minded physicists - and DEFINITELY NOT to non-scientifically minded common sense people such as Mrs. Gratke.

I believe Einstein's second quote says that to realize fundamental new truths one must be prepared to rise above the unreasonable constraints of blind conformity enforced by unreceptive conformists. Einstein experienced moderate opposition at various times throughout his life, but he also experienced some instances of *'violent opposition'*. Einstein's stressing of the idea of 'violent opposition' might at first seem to be an exaggeration, but consider just a few experiences of others.

- Nicholas Copernicus delayed the publication of his famous sun centered astronomy until he lay on his deathbed in 1543. He did this so that he was beyond any possible reprisal from the dreaded 'church inquisitors'.

- For subsequently proclaiming occultist ideas, such as the Copernican view that the earth went around the sun, the philosopher Giodarno Bruno was burned at the stake in 1600.
- For supporting the Copernican idea of a sun-centered solar system, Galileo Galilei was reminded about 'the instruments of interrogation'. Therefore, he publicly recanted his statements. Even with this renouncement, Galileo lived the last 11 years of his life under house arrest until his death in 1642.
- In his great work the "PRINCIPIA" in 1687, Isaac Newton proposed his universal law of gravity, but he withheld any possible explanation of the cause of gravity. Newton literally feared any description of causality might be misinterpreted as a proclamation of occultic powers. When he later felt the times were more tolerant, Newton did offer his vision for an underlying physical causality of gravity.

In the medieval era, clearly one needed to be careful about offering bold new scientific ideas. You may think that we are well past such uncivilized primitive retribution, but that was not the case for Einstein. As a Jewish physicist in Nazi Germany proposing radical new ideas, Einstein was a very vulnerable target. His ideas were not just denounced, they were derided as "*Inferior Jew Science*". Totally unprecedented within public scientific forums, Einstein was heckled and harassed by organized opponents for his views about Relativity. He was even threatened with death by nameless antagonists. This was no idle threat as other prominent Jews had been attacked and killed in those turbulent times in Germany.

Einstein learned a cold reality that with hatred and bitterness in their hearts, those with 'mediocre minds' are prepared to treat proponents of new ideas in an evil way. New thinkers are often vilified because their new ideas are too troublesome and must literally be forbidden. Newton's eventual views of physical causality underlying gravity were denounced by the scientists of his era.

Likewise, Einstein's views of physical causality underlying Quantum Mechanics were also denounced by his peers. In the extreme, proponents such as Bruno were killed for expressing their thoughts, and Galileo and Einstein had been threatened with death. In fear of extreme reaction to their new ideas, proponents such as Copernicus and Newton delayed announcements of all they envisioned.

Unjust intolerance emerges when controlling authorities outgrow the natural restraint of humility and common sense. The zeal of 'mediocre minded' guardians simply opposes productive nonconformity and legitimate change. Once a flaw in thinking has become ingrained, it becomes difficult to correct. A flawed thought is not upheld by clear thinking but by non-thinking adherence to authoritarian dogma.

Today, most of us probably view the centralized thought control of Hitler's Nazi Germany or that of Stalin's Communist USSR as aberrations. Surely in today's world we should expect a much more open consideration of new ideas. While we should expect today's opposition to new ideas to be more civilized, pressures are nevertheless brought to bear. The psychology and sociology of science is itself an interesting subject of study, and is discussed in the *Full-Book Edition*. In the meantime, let us consider that Einstein had also experienced rejection in our civilized modern world outside of Nazi Germany. Einstein's vision of unification was opposed by prominent scientists who were advancing some very weird ideas in the form of 'Quantum Mechanics'. The intolerance of Quantum Theorists may have been civilized, but nevertheless unyielding and intellectually shocking.

Einstein's quest for his 'Unified Field Theory' seemed to be hopelessly positioned outside the boundaries established for Modern Physics. These strange boundaries encompass the ideas of a nearly infinite number of particles that can instantaneously communicate with each other no matter how great their separation. This 'instantaneous non-local entanglement' mystically connects particles via an 'instantaneous non-local causality'. These particles are themselves dissolved and exist as non-physical 'quantum probability waves' that are distributed throughout an infinite number of 'parallel universes'. Each 'parallel universe' is then comprised of at least several 'extra dimensions'. This appears to be nothing less than 'anti-physical' and therefore mystical. One book title expresses this idea of mysticism quite well, "*THE TAO OF PHYSICS*". For opposing this view, Einstein was regarded as the old man who could not keep up with the great revelations of Modern Physics. The *'mediocre minds'* could not accept nor tolerate Einstein's 'simple minded' hidden physicality.

Newton and Einstein, the two greatest physicists of all time, both thought in terms of an underlying physicality. Specifically, I believe each labored under the common sense perception that:

- Physical objects possess the quality of '*Objective Physical Existence*'.
- Physical objects we observe dwell within our '*Single Universe*' of '*Three Spacial Dimensions*'.
- Physical objects are governed by a '*Principle of Local Physical Causality*'.

Even though Newton and Einstein are the two greatest physicists of all time, their most basic sense of physicality has been rejected. Essential aspects of both Newton's and Einstein's visions about *Objective Physicality* were unjustly rejected by their respective peers and also by today's physicists. I personally support and am encouraged by the thoughtful intent, the insightful spirit, and the scientific philosophy of their respective visions. I especially appreciate the dilemmas of Newton and Einstein, because my own vision of a *Unified Physical Field Theory* will likewise be confronted with unjust opposition.

Einstein's warning about *'opposition from mediocre minds'* is a testimony to his wisdom, which was tempered by the trials and tribulations of his amazing life.

In the book "*The Expanded QUOTABLE EINSTEIN*", I just recently came across the continuation of the 'Great Spirits' quote.

"The mediocre mind is incapable of understanding the man who refuses to bow blindly to conventional prejudices and chooses instead to express his opinions courageously and honestly."

In that same book, Einstein is reported to say,

> ***"A foolish faith in authority***
> ***is the worst enemy of truth."***

Prologue - Article 4

Einstein's Vision - Revising the Legacy

Albert Einstein was born March 14, 1879 and died April 18, 1955. The year 1905 became the miracle year for this patent clerk in Bern Switzerland. As part of that miracle he unveiled his 'Special Theory of Relativity' (the relativity of non-accelerated motion), and also $E=mc^2$ (the idea that mass and energy are equivalent). In 1916, Einstein unveiled his 'General Theory of Relativity' (the relativity of gravity and accelerated motion), and also 'curved spacetime' (the four dimensional non-Euclidean geometry of space and time). Einstein's legacy of Relativity is generally held up as Einstein's greatest contribution to science. However, this assessment is 'NOT QUITE' correct. It seems that everyone, including physicists, misperceive Einstein in some way. The following ten misperceptions, though not exhaustive, serve to frame the scope of misunderstandings.

Misperception #1: Many in the general public are surprised to learn that Einstein received his Noble Prize in 1921 for discovering the 'photoelectric effect', (a quantum phenomenon). As part of his miracle year of 1905, Einstein revealed that light formed tiny quantized bundles of energy called 'photons', which interact with matter via the photoelectric effect. The energies of these light quanta are specified by the companion energy equation $E=hf$ (an equation originated by Max Planck). Thus, light acted both like a particle and a wave in a paradoxical manner called 'wave-particle duality'. The Noble Prize citation made no explicit mention of Einstein's work with Relativity, which somewhat annoyed him. Einstein responded with subtle humor. In his acceptance speech for the Noble Prize, Einstein spoke only about Relativity and said nothing about the photoelectric effect.

Misperception #2: Most in the general public are also unaware that in 1905 Einstein wrote four 'seminal' papers (important, original and pioneering). Any one of these four papers would have made him a notable figure in the history of physics. Einstein presented two papers in 1905 on Relativity, (Special Relativity and $E=mc^2$), and one on Quantum Mechanics, (the photoelectric effect with $E=hf$). The fourth paper dealt with 'Brownian motion', (an atomic phenomena of classical thermodynamics). By explaining the jiggling of tiny visible particles as being induced by collisions with the even more tiny invisible particles of a fluid, Einstein confirmed to doubting scientists the kinetic theory of molecules. Einstein thereby provided unique credibility for the concept of atoms. The year 1905 was truly Einstein's miracle year of published revelations.

Misperception #3: Einstein stated his biggest mistake was proposing the 'cosmological constant', (a fudge factor in his 'gravitational field equation' to keep the universe from collapsing). Today, it is not clear that the cosmological constant was a mistake. Furthermore, (as will be noted in Misperception #10), I contend that Einstein became involved with 26 other mistakes about Relativity. The combined significance of those other mistakes is far greater than the possible error of the cosmological constant.

Misperception #4: Einstein is quoted as saying, *"Do not worry about your difficulties in mathematics; I assure you that mine are still greater."* It seems incredible that a genius, such as Einstein, would have difficulties with mathematics, but that is far more significant than has been imagined. Almost everyone is surprised to learn that this very difficulty with mathematics undercut the visions that Einstein sought to convey about Relativity, Quantum Mechanics and unification. One might also imagine in discussing his difficulties with mathematics, Einstein was not just speaking for himself. He might also have been addressing a culture of misplaced emphasis upon mathematics by the whole community of Modern Physicists. The more significant point is that the mathematics did not exactly mirror his vision of *Physical Reality*. The mathematics seemed to loose something in translation and thereby seemed to suppress the full expression of his inner vision.

Misperception #5: In the 1920s, as the ideas of Quantum Mechanics drifted into the realm of non-physical absurdity, Einstein sought to reassert an underlying physicality as the proper basis of science. He argued valiantly, yet apparently in vain, against the non-physical weirdness that had crept into Quantum Mechanics. For his efforts, he was ridiculed by his contemporaries. Einstein, at the peak of his career in the late 1920s, was lambasted and considered a stubborn old man for not accepting the new 'pseudo-science' of Quantum Mechanics. Contrary to common belief today, I claim Einstein 'WAS RIGHT' about Quantum Mechanics, but his arguments were 'NOT QUITE' sufficient to convince the scientific community. Sadly, Modern Physics was not only embracing non-physicality, it was falling into a disgraceful 'sophistry of anti-physicality'. This sophistry and its disgrace have persisted ever since.

Misperception #6: During the later part of his life, (from the 1920s to his death in 1955), Einstein championed the idea of a 'Unified Field Theory'. Einstein sought a way to overcome the 'sophistry of anti-physicality' by trying to discover an underlying *Objective Physical Reality.* He envisioned that distortions of a single field produced all the elementary forces, and that disturbances of this single field could also form what we perceive as all the elementary particles. Though Einstein never succeeded in this effort, I claim Einstein's vision for unification was essentially correct. The mathematical abstractions of Quantum Mechanics and of Einstein's own Relativity stymied Einstein's quest. Contrary to common belief today, Einstein 'WAS RIGHT' about unification. However, Einstein was 'NOT QUITE' able to overcome his own abstract formulations of Relativity and his own abstract ideas about fields.

Misperception #7: As historical figures progress through their lives, their thoughts can evolve, but historians may simply fail to properly report these changes. Einstein initially reveled in Ernst Mach's philosophy of relativism, but historians often fail to report that late in his life, Einstein rejected Mach's philosophy. Mach asserted that objects sensed the combined presence of all the stars and matter of the universe and thereby were able to manifest a sense of acceleration. Einstein finally wrote, *"Of Mach, we should speak no more."* I claim Einstein 'WAS RIGHT' in finally rejecting Mach's philosophy. However, Einstein was so weak and tardy in his renouncement that he did 'NOT QUITE' convince all the historians that he really meant it.

Misperception #8: As disciples put ideas of a master intellect into their own framework of thinking, they add their own interpretation. Historians often falsely attribute these interpretations to the master. Einstein did not like the concept of 'blackholes' because he thought they were un-physical. Yet Einstein's name is often erroneously attached to the idea of 'blackholes' without any mention of his strong renunciation of these 'dark stars'. If his rejection of 'blackholes' is mentioned at all, it is typically passed off as the delirium of an old man who was once a genius. I claim Einstein 'WAS RIGHT' in rejecting 'blackholes' (which are to be replaced by *Greyholes*). However, Einstein could 'NOT QUITE' take advantage of his physical intuition to develop a winning argument against 'blackholes'. This is where Einstein missed a most unique and most important opportunity.

Misperception #9: A master may take a carefully qualified and nuanced position, which disciples and historians may try to compress into a brief distorted slogan. Here is a most regrettable distortion: "*The speed of light is constant!*" Einstein advanced the idea that the 'measured' speed of light is constant, while also showing that light slowed down in a gravitational field. Recognizing this nuance about the speed of light is to behold the key for discerning the true nature of curved spacetime, for exposing the fallacy of 'blackholes', and for 'Perfecting $E=mc^2$'. Even Einstein, playing with coyness, became entangled by his own subtlety.

Misperception #10: Einstein has been revered as the genius who discovered and championed the counter-intuitive ideas of Relativity, but who could 'NOT QUITE' come to grips with Quantum Mechanics and unification. However, I believe just the reverse is true. Einstein 'WAS RIGHT' in concept about Quantum Mechanics and unification, but was 'NOT QUITE' right about Relativity. While Einstein did have a sense of physicality for Relativity, he did not openly promote underlying physicalities. Einstein thereby fell into a trap of his own making. Einstein unwittingly promoted and/or tolerated 26 mistakes about Relativity, which then undercut his own visions of physicality for Relativity, Quantum Mechanics, and unification. (The *Full-Book Edition* will discuss these 26 mistakes in detail.)

Let us take care to note that some of those 26 mistakes about Relativity are NOT mistakes originated by Einstein. However, those mistakes not originated by Einstein were either indirectly facilitated by Einstein or not opposed strongly enough by Einstein. In some cases, Einstein's coyness was misleading. In other cases, even Einstein was overwhelmed by the snowballing enthusiasm of others for the abstractions of Relativity. Overall, Einstein was guided by a strong sense of physicality, but his formal mathematical presentations disguised and curtailed this critically important quality. Quite simply, Einstein's presentation of Relativity seemed to undercut the foundation of *Common Sense Physicality*. Einstein's Relativity set a tone of 'detached observationalism' that then mushroomed into the 'sophistry of anti-physicality' that has now come to disgrace Modern Physics.

By not properly reporting the subtleties of Einstein's ideas, the general public and most physicists in some way are led to misperceive Einstein and his ideas. Overall, Einstein's reputation should be slightly diminished for Relativity but greatly elevated for Quantum Mechanics and unification. With these offsetting shifts in assessment, Einstein's legacy should remain exceptionally high. Albert Einstein is truly one of the greatest geniuses of all time!

It is my view that,

In the prior history of Physics,
Albert Einstein is exceeded only by
the incomparable Sir Isaac Newton.

Prologue - Article 5

Objective Common-Sense Physical-Reality

To restore the *Objective Physical Reality* which Einstein sought, we must debunk the non-physical myths of Modern Physics. We must unify all natural phenomena within an *Objective Physical Universe*. We must explain confirmed observations based upon *Localized Physicality* of physical variables within today's three-dimensional universe. This is the essence and the goal of *Common Sense Physicality.*

Unification via a plethora of particles, or only during the big bang era, or only in extra spacial dimensions are all oxymoronic in that they do NOT unify in the here and now! Unification theories outside of *Objective Physical Reality* are just modern day renditions of ancient Greek epicycles (like Don Quixote charging at windmills). The physics of today is blind to *Objective Physicality* because it is duped by non-physical premises, such as 'parallel ghostly universes', 'instantaneous non-local causality' and 'extra-dimensionality'. These are the very antithesis of *Common Sense Physicality*; and they simply can NOT be correct!

Misrepresenting the essence of *Localized Physicality* has caused quantum wave effects and EPR experiments to be misinterpreted. Quantum wave phenomena do NOT justify 'parallel ghostly universes'; and the evidence of EPR experiments does NOT confirm 'instantaneous non-local causality'! String theory misrepresents our universe as being an 'extra-dimensional Hyperspace'. Time and space do NOT curve into any 'extra dimension'. Two relatively moving clocks do NOT both run slower than each other. The vacuum of space along with forces and structures at all levels, from sub-atomic to cosmological scales, have been systematically misrepresented. Elementary particles, atoms, 'black-holes', quasars, and dark matter are misunderstood. Underneath all of these lie deeper descriptions based upon a *Localized Objective Physicality.*

We exist in the here and now, not somewhere else, nor at some other time. We do not dwell in parallel universes, nor in extra spacial dimensions, and neither in the past, nor in the future. Instead of extra dimensions, we should consider additional variables within our single three-dimensional universe. Variables such as various parameters of shape, size, orientation, position, motion, compression, stress, and strain represent viable aspects of *Common Sense Physicality*. To overcome the fallacies of modern sophistry, we shall explain known observations and all natural phenomena based upon *Local Causality* of physical variables within today's single universe of just three spacial dimensions.

Einstein objected to the 'non-physical' and 'anti-physical' developments of Modern Physics. However, Einstein did contribute to this misleading process by helping to establish the approach of 'detached observationalism' for his Theories of Relativity. Einstein was trying to say what we should expect to observe and avoided discussing that which we could not directly observe. Modern Physics then expanded 'detached observationalism' within Quantum Mechanics but in a different way. The Theory of Quantum Mechanics evolved to describe what we cannot observe as existing in some 'fuzzy wuzzy' state of 'non-physicality'. This self-deception impugns and decimates the concept of *Common Sense Physicality*. This 'way of thinking' has led us to:

- Develop distorted theoretical models.
- Invoke obfuscating mathematics.
- Systematically misinterpret key observations.
- Extrapolate false precision.
- Conjure up fanciful non-existent concoctions.

Einstein objected to this 'non-physical' and 'anti-physical' development for Modern Physics. Einstein stood virtually alone as a beacon of hope for *Common Sense Physicality* in a darkness of semi-mysticism parading itself as science. If we must describe objects when we are not observing them at any given moment, Einstein felt that we should ascribe to them the qualities of *Objective Physical Existence*. I agree whole heartily with Einstein about this perception of a 'hidden underlying physicality'. This kind of physicality is *Objective Common Sense Physical Reality*.

Given the current state of Modern Physics, we are smugly advised to abandon *Common Sense Physicality* because it is naive and inadequate for describing how our universe works. Einstein and a few others have disagreed with this vision of 'anti-physicality', and I proudly join this group of dissidents. Yes, our common sense has deceived us, but herein, we now seek to revitalize our collective common sense. Meanwhile, the insulters of *Common Sense Physicality* fail to realize the universe really is structured with an underlying *Objective Physical Reality.* Thus, both 'naive proponents' and 'smug insulters' of *Common Sense Physicality* need to let their respective ways of thinking evolve.

Much of Modern Physics is correct, but that part, which is incorrect, is profoundly fundamental. In touting a flawed foundation, Modern Physics promotes foolish non-physical descriptions of reality, insults our common sense intelligence, and undercuts the dignity of humanity's noble quest for truth. Misled by the fallacies of modern orthodoxies, valuable resources are sometimes diverted into un-productive research and legitimate advancements are sometimes obstructed. Of course, the most valuable resource to be squandered is the collection of talented young minds that become misdirected. The time has now come to correct this misadventure and to return to the pathway of *Objective Common Sense Physical Reality.*

Prologue - Article 6

Back Through the Looking Glass

In the classical children's story "*ALICE IN WONDERLAND*", Alice steps through a 'looking glass' (a mirror) into a 'world of make-believe' where everything seems absurdly different. Somehow, in the early 1900s, we fell through a looking glass, formed by the $E=mc^2$, which is the 'self-energy' of matter; by the $E=hf$, which is the energy of 'wave-particle duality' for light; and especially by 'Extreme Quantum Weirdness', which is the antithesis of *Physical Reality.* We appeared to take a one-way trip into

Modern Physics causing us to become trapped in a crazy 'Wonderland' sprinkled with make-believe absurdities.

In his miracle year of 1905, Einstein presented both E=hf for the photoelectric energy of photons and $E=mc^2$ for the energy of rest-mass particles. The photoelectric effect deals with the energy of photons, the energy quanta of light particles. The equation $E=mc^2$ deals with the energy of ordinary matter and should have been related to the energy quanta of rest-mass particles. The ideas behind these two energies were not fully developed. Together, the distorted aspects of E=hf and $E=mc^2$ and the failure to connect $E=mc^2$ to the energy of quantum states helped to transport us into the bizarre 'Wonderland' of Modern Physics. Nevertheless, it is amazing that Einstein (as a lone researcher) became involved in an original way with both E=hf and $E=mc^2$ in the very same year - his miracle year of 1905.

The critical point is that Einstein's vision of $E=mc^2$ for the self-energy of matter and E=hf for the wave-particle duality self-energy of light were incomplete. Being incomplete about these basic energies weakened his ability to fend off the subsequent 'anti-physical' pronouncements. Einstein was unable to stave off the miscues initiated by his own General Relativity in 1916 and the gross absurdities of 'Extreme Quantum Weirdness' introduced in 1926. We thus ended up in the 'Wonderland' of make-believe called Modern Physics.

By 'Perfecting $E=mc^2$', by 'Clarifying E=hf', and by 'Debunking Quantum Weirdness' we can re-open the looking glass. We can thus establish a pathway to escape the 'Wonderland' and to return to the 'Real World' - to return to *Objective Physical Reality.*

$E=mc^2$ equates energy and mass, where mass cannot only be converted into energy, but mass itself is energy. Einstein revealed motion caused mass to change and to thereby manifest kinetic energy. Thus, the mass in $E=mc^2$ was understood to be the actual repository of the classical energy of motion for ordinary matter. I shall be revealing the unique way that $E=mc^2$ is also induced to change by a gravitational field. Thus, $E=mc^2$ can now be understood to also be the actual repository of the classical potential energy of gravity. I contend the rest-energy of $E=mc^2$ is reduced in a gravitational field. Thus, for a rest-mass particle free-falling in a gravitational field, the lost rest-energy is converted into the kinetic energy portion of $E=mc^2$. To the extent that rest-energy portion of $E=mc^2$ is reduced by gravity, gravitational potential energy is being released.

Quite simply, Einstein's 'General Relativity' ignores how gravity alters $E=mc^2$ and falsely implies that the total energy of a particle can become negative. Einstein employed the classical expression for the potential energy of gravity, but this is only an approximation that can improperly generate excessive negative energy. Below the 'supposed' 'horizon' of a 'blackhole', the excessive negative potential energy is falsely imagined to produce negative total energy for a particle. I proclaim that the total energy of ordinary matter always remains positive, and the 'horizon' of a 'blackhole' never forms! The mathematics employed to incorrectly proclaim the formation of 'blackholes' has simply misrepresented *Physical Reality*.

Whereas Einstein did not complete $E=mc^2$ for gravity, today's Physicists go too far in assuming how gravity effects the energy of light particles. They erroneously insist gravity alters the $E=hf$ energies of photons in transmission. If gravity reduces escaping photon energies enough, photons 'supposedly' do not escape gravity, but fall back. The maximum height of the returning photons would constitute a 'blackhole horizon'. I proclaim that the energy of photons is NOT altered by transmission through a gravitational field. Gravitationally redshifted photon energies are altered at the moment of emission due to altered $E=mc^2$ energies of their emitting source within a gravitational field.

Together, the misunderstandings about $E=mc^2$ and $E=hf$ led to the fallacious conjuring of 'blackholes'. Einstein stated that he did not believe in 'blackholes' and that gravity altered the frequency of escaping photons at the moment of their emission, but he failed to act upon these correct beliefs. I advise we consider *Greyholes* as the replacement for the very dense objects previously imagined to be 'blackholes'. More importantly though, the continuous variation of the quantum state energy along with a rest-mass particle's other parameters is the *Objective Physical Reality* that unifies Quantum Mechanics and General Relativity.

As we correct the misconceptions about $E=mc^2$ and E=hf, we shall begin to unravel the Objective *Physical Realties* underlying *Unified Physicality.*

- The changes of kinetic and potential energy within $E=mc^2$ correspond to changes in the quantum state harmonic energy of rest-mass particle formation.
- Physical waves of a physical medium of space form the elementary particles with a wave sub-structure that underlies wave-particle duality.
- The quantized wave harmonic energy of an $E=mc^2$ particle can mutate to another harmonic state by absorbing or emitting another particle.
- The quantized wave harmonic energy of $E=mc^2$ can also smoothly vary along with a rest-mass particle's mass, internal timing, size and shape.
- These smooth harmonic variations occur with absolute motion and also with position within a gravitational field to underlie Special and General Reality.

Of Einstein's miscues about Relativity, those concerning $E=mc^2$ are the most significant. I contend that Einstein's biggest errors were his failing to properly extend $E=mc^2$ into the realm of non-Euclidean Spacetime and Quantum Mechanics. When we 'Perfect $E=mc^2$' we start to unlock the mystery haunting the great quest for *Unified Physicality.* By 'Perfecting $E=mc^2$', we can begin to reformulate General Relativity and Quantum Mechanics within a context of *Objective Physicality.* This then forms a basis of an *Objective Quantum Relativity* and constitutes an underlying essence for my *Unified Physical Field Theory.*

By discovering the *Physical Reality* underlying $E=mc^2$ and E=hf, and by 'Debunking Quantum Weirdness', we can open a return pathway through 'Alice's looking glass'. We can return to the 'Real World' - a world based upon *Objective Common Sense Physical Reality.* While the year 1905 became known as Einstein's miracle year, it actually opened a one-way path for the 'looking glass' of Modern Physics. Then the General Relativity of 1916 and the Quantum Mechanics of 1926 pushed us through that 'looking glass' into a 'Wonderland' accented with strange 'anti-physicalities'.

Prologue - Article 7

Reconciling Detached Observationalism

Discerning *Objective Physical Reality* is our primary objective. However, let there also be an implicit understanding that this physicality must eventually be reconciled to mathematical formalisms. I seek to decipher the confusing mathematical abstractions created by 'detached observationalism'. I plan to reattach the 'detached observationalism' of Modern Physics to an actual underlying *Physical Reality.* A mathematically minded physicist would normally like to lead with the mathematical presentations, but herein I am leading with the physical description.

Even though Einstein was the greatest opponent to 'Extreme Quantum Weirdness', he was the instigator of 'detached observationalism' within Modern Physics. The very nature of Einstein's theories of Relativity promoted the idea of "*describing observation without describing underlying physicality*", which I call 'detached observationalism'. The observations can be correct, but if we do not understand the common sense of its underlying physicality, then we are left in a state of unresolved perception. The resulting confusion creates an aura of mystery, which unavoidably leads to perceptions of 'anti-physicality'. By 'reconciling detached observationalism', we can recover our good sense for the underlying workings of all natural physical phenomena.

There are three main bodies of mathematical formulation within modern physics, which we will eventually seek to tie together. The first body of mathematical formalism involves Einstein's Relativity. The second body of mathematical formalism involves Quantum Mechanics. A third body of emerging mathematical formalism involves String Theory. Except for $E=mc^2$ and $E=hf$, we will see little mathematics in this *Monograph.* What we shall see is a description of the underlying physicality, which will allow the basis of the mathematical connections to be understood in concept.

From a physicist's point of view, the goal of this book is to describe the plausibility of an underlying physicality for all natural phenomena. However, to become involved with the mathematics in a detailed way at this time would detract from the presentation of that underlying physicality. Nevertheless, to satisfy some reviewers and some avid readers, I am relenting by placing some key mathematical sequences in an *Appendix* in the *Full-Book Edition*, which the more hearty can pursue if they wish. If you are not so inclined, do not feel badly, but do take heart in knowing that such detail will be available.

Keep in mind I seek to describe nothing less than the underlying unifying physical structure of all natural phenomena. I seek to unify the Relativity, the Quantum Mechanics, and the String Theory of Modern Physics. I seek to unify the microscopic formation of particles with the macroscopic formation of large-scale cosmological structures.

With this vision of *Unified Physicality* in mind, let us now contemplate my *Unified Physical Field Theory*. Let us consider the basis upon which we can eventually re-attach the 'detached observationalism' of Modern Physics to an *Objective Physical Reality*. Let us thus witness the unveiling of the year 2005 as "*THE NEW MIRACLE YEAR*".

Einstein did not like the 'anti-physical' aspects of Modern Physics and openly proclaimed his disbelief. I salute Einstein's independence, genius, and fortitude in keeping the faith for *Objective Physical Reality*. I most strongly agree with Einstein in rejecting the 'anti-physicalities' of 'Extreme Quantum Weirdness'. I thank him for his inspiration of a 'Unified Field Theory'. Now in 2005, I propose that we step 'back through the looking glass' into the 'Real World' governed by *Physical Reality* and *Ultimate Unification*.

Prologue Post Script: This brief *Prologue* has sought to pay proper tribute to Albert Einstein and to set the stage for building upon his legacy. Einstein had a vision for Relativity, Quantum Mechanics, and *Ultimate Unification*, and that vision was essentially correct but incomplete. I now seek to advance that vision of *Ultimate Unification.* The most fundamental reality upon which I seek to build involves his fantastic energy equation $E=mc^2$. In *Chapter 1*, I will formally introduce my version of this equation as a step toward 'Perfecting $E=mc^2$'.

Einstein's Version: $$E = mc^2 = m_o c^2 \gamma$$

Gratke's Version: $$E = mc^2 = m_o c_o^2 \gamma_M^1 \gamma_G^{-1}$$

Einstein strongly objected to the 'anti-physicalities' of Quantum Mechanics and sought to repudiate those aspects of it that were most weird. I seek to vindicate his unappreciated vigil of standing alone against the physical absurdities of Quantum Mechanics. In *Chapter 1*, I seek to 'Debunk Quantum Weirdness' and to remove this major obstacle for the development of a *Common Sense* vision of *Unified Physicality.*

I believe Albert's 'Great Spirit' will smile down upon this work. In one sense the smile might be sheepish for having missed completing $E=mc^2$. In another sense the smile will be of deep satisfaction for having anticipated the correction of Quantum Mechanics and for wisely envisioning a 'Unified Field Theory'.

Chapter 1

The Vision of Unified Physicality

Prepare to witness the unveiling of 'A New Miracle Year in Physics'.

In 1989, Albert Einstein's dream of unification had become my dream, and now I believe *Unified Physicality* has become my reality. As you read onward, this dream could also become your reality. This reality is not just a state of mind. This reality is objective. It exists regardless of our contemplation of it. It makes good sense. We can actually understand it within our perceptions of common sense. This vision of *Unified Physicality* can thus be said to recognize the *'Objective Common Sense Physical Reality'* that actually exists and has so long awaited our discovery.

Chapter Article 1.1

Common Sense Physicality

This conquest of *Unified Physicality* cannot be timid. It requires the bold renouncement of those significant errors that obstruct *Common Sense Physicality.* I herein seek to expose those errors and to present a new vision of *Objective Physicality.* Average non-scientific readers may become pleasantly surprised. You will come to discover you can actually understand these ideas of *Unified Physicality.* You will come to appreciate how all natural phenomena can indeed occur within a context of a '*Objective Common Sense Physical Reality'*.

I consider the following expressions to be virtually synonymous, and use them interchangeably:

- *'Objective Common Sense Physical Reality'*
- *'Objective Common Sense Physicality'*
- *'Common Sense Physicality'*
- *'Objective Physical Reality'*
- *'Objective Physicality'*
- *'Physical Reality'*

This physicality is not easily discerned, and it is not simplistic, obvious, or common. in fact, it is partially hidden from our direct observation, so it has been quite difficult to recognize.

Regardless of how unlikely an initial facet of *Objective Physicality* may appear, once fully explained, it should be understandable within the perspective of our common sense. The scope of our common sense may need to grow, but ultimately our common sense should find comfort with the plausible form of *Objective Physical Existence* being described. Surprisingly, this is NOT the case with Modern Physics in its current form. After the explanations of Modern Physics are given, its form still appears to be implausible and contrary to common sense. Today's physics thus exudes a sense of mystery, 'anti-physicality', and even 'scientific mysticism'.

I define *'Objective Common Sense Physical Reality'* as "*empirically-confirmed logical-physicality*", which encompasses the three following points of physicality:

- To possess *Objective Physicality* an entity must exude innate qualities of 'physical existence' and 'physical persistence' independent of our thoughts.
- For a proposition of *Objective Physicality* to be viable, it must be consistent with known observations to demonstrate '*empirical confirmation*'.

- *Objective Physicality* evolves with time in a connected causal way within its immediate surroundings to demonstrate '*logical physicality*'.

The main problem with Modern Physics is that it has erroneously 'discovered' that *Common Sense Physicality* is incompatible with known observations. Thus, it falsely concluded that *Objective Physicality* is INCAPABLE of representing reality. Consequently, today's physics tends to view all proponents of *Objective Common Sense Physicality* as being naive. Indeed, many (but not all) common sense proponents are naive. However, as you read on, you will have a chance to judge for yourself the viability of *Common Sense Physicality*.

I believe today's physicists have allowed themselves to be misled, and thereby conjure up relationships for physics that are NOT physical. I take exception to key abstract proclamations of Modern Physics. Employing my vision of *Objective Physicality*, I propose to demonstrate that the following 'anti-physical' ideas are simply wrong. It will sound harsh and boastful, but to fix this problem, I must point out what I believe is WRONG?

- I believe ***Extra*** 'dimensions' for curved spacetime, Strings, and Parallel Existences are WRONG!
- I believe ***Non-Physical*** 'quantum probability waves' are WRONG!
- I believe ***Dissolved*** 'parallel existences' are WRONG!
- I believe ***Ghostly*** 'parallel universes' are WRONG!
- I believe ***Instantaneous*** 'non-local quantum entanglement' is WRONG!
- I believe ***Instantaneous*** 'non-local quantum causality' is WRONG!
- I believe ***Infinite Intensity*** 'singularities' for point particles and 'blackholes' are WRONG!

It is one thing to use mathematics to predict new facets of *Objective Physicality* which we have not yet observed. That is a proper role of applied mathematics within theoretical science. However, I contend it is quite improper for physicists to misuse mathematics to predict non-physical relationships. I claim Modern Physics has succumbed to using 'obfuscating mathematics' to promote strange non-physical propositions that are mysterious, 'anti-physical', and outright mystical. The misapplication of mathematics thus obscures our ability to perceive the actual *Physical Reality*, which we desperately seek to know. We should expect the application of mathematics within physics to be confined to describing plausibility's that exude the properties of *Objective Physicality.* Einstein once said,

"As far as the laws of mathematics refer to reality, they are not certain; and as far as they are certain, they do not refer to reality."

Modern Physics has become encumbered by a 'sophistry of anti-physicality' that blinds our collective judgement. My dictionary defines sophistry and its derivatives as follows:

- ***Sophistry*** is *"reasoning that is superficially plausible but is actually fallacious"*.
- ***Sophism*** is "*an argument correct in form or appearance but actually invalid*".
- ***Sophist*** is "*a captious or fallacious reasoner*".

These definitions all stem from the Ancient Greek Sophists who were noted for "*their adroit and subtle skills of argumentation that were often tainted with specious reasoning*".

This language about sophistry must seem to be a rather damning indictment, but consider just this one example of a very strange pronouncement. The moon 'supposedly' ceases to exist as we know it when no one is looking at it, because it is only our observation of the moon that 'supposedly' makes it exist as we normally see it. This version of Quantum Mechanics promotes the silly idea of 'thought controlled existence' - more precisely, 'physical non-existence' or 'non-physical existence'.

Many readers might think I am misreporting this story about the moon dissolving when no one looks at it. Let me reassure you that this is what Modern Quantum Theory says. Amazingly, this Modern Quantum Theory promotes the idea of 'non-local' 'thought-controlled' 'mystical-existence'. This strange existence is 'instantaneously-entangled' throughout an 'infinite-number' of 'extra-dimensional' 'ghostly' 'parallel-universes'. I contend this is blatant 'anti-physical mysticism'! This is the 'sophistry of anti-physicality' that undermines the integrity of Modern Physics.

Albert Einstein strongly objected to this formulation of Quantum Mechanics, and I agree with him. The breakdown in logic misleading Quantum Theory has ensued because *Common Sense Physicality* was prematurely rejected. Quantum Theory misapplies mathematics because we could not see how any version of *Common Sense Physicality* could possibly produce the effects we actually observe. I propose to show that observed quantum effects are compatible with a vision of *Objective Physical Reality.*

In my vision of *Physical Reality*, at least the moon maintains its *Objective Physical Existence* regardless of our collective gazing upon it. Once we overcome the 'sophistry of anti-physicality' that has encumbered Modern Physics, we can then begin to develop a viable theory of *Unified Physicality.* In direct conflict with the current tenets of Modern Physics, this *Unified Physicality* is based upon a vision of *Objective Common Sense Physicality.*

It is always wise to be prudent about any bold new pronouncements. However, in this case, physicists will certainly be much more reluctant than mere caution would suggest. Since their very training encourages them to accept flawed 'anti-physical' premises, they have been conditioned to be biased against *Objective Physicality.* To overcome this bias, I seek to expose the specific fallacies that obstruct the proper discernment of *Physical Reality.* I seek to replace the fallacious 'anti-physical' propositions lurking within Modern Physics with my own visions of *Physical Reality*!

Many fallacies, nuanced caveats, and technical details will be explored. Most readers will come to understand this vision of *Common Sense Physicality* because it is quite natural. It actually does make good sense. Once you realize that you can throw off the most absurd aspects of Modern Theoretical Physics, you will discover you too can comprehend *Unified Physicality.*

Chapter Article 1.2

The Postulate of Physical Unification

I believe my perception of *Common Sense Physicality* and my vision for the *Unified Physical Field Theory* can be justified. To do so, we must first be prepared to accept my basic *Postulate of Physical Unification.* For purposes of clarity and emphasis, I present this postulate in eight parts as follows:

- All the natural phenomena we observe possess an *Objective Physicality* that dwells within our single universe of only three spacial dimensions.
- This *Objective Physicality* in three-dimensional space changes in time, and time evolves in only one-way, from past to present to future.
- The *Cosmic Continuum* is the physical medium of space, and it is the sole physical substance in our universe. It physically fills and forms our universe.
- Non-linear distortions of the *Cosmic Continuum* form a mathematically rich *Physical Field* underlying all the natural phenomena that we observe.
- The 'physical existence' and 'physical persistence' of the *Physical Field* constitutes the first two qualities of physicality inherent in *Objective Physicality.*
- The qualities of 'physical existence' and 'physical persistence' are sustained independent of our observation of *Objective Physicality.*
- The 'localized physical interconnectedness' of the *Physical Field* constitutes the third quality of physicality inherent in *Objective Physicality.*
- The very qualities of *Objective Physicality* require all natural phenomena to be governed by a *Principle of Local Physical Causality.*

This *Postulate of Physical Unification* describes my perception of the *Objective Physicality* underlying my *Unified Physical Field Theory.* I specifically wish to emphasize the basic character of causality. The *Principle of Local Physical Causality* dictates that any effect communicated over any distance requires both an *Objective Physical* connection and some amount of time to communicate the effect. The idea of *Objective Physicality* with *Local Physical Causality* functioning within our single universe of just three spacial dimensions constitutes the essence for my idea of *Common Sense Physicality.*

Here is the essential challenge. This vision of *Common Sense Physicality* directly conflicts with the most basic tenets of Modern Physics. I consider the current underlying tenets of physics, as they are currently presented, to be 'anti-physical' and therefore flawed. I believe these flawed tenets have impaired our collective thinking about underlying physicalities and have undercut the very integrity of science. Herein, I propose to clearly expose what I consider to be the primary errors inherent in those basic tenets of Modern Physics.

In the process of structuring a logical argument or a particular theory of science, one must start with an agreed upon set of premises. In general, a postulate is a premise, and a postulate cannot be proved. Since my premise is a postulation, I cannot prove it. To consider my overall proposition of unification, please examine and consider the plausibility of my *Postulate of Physical Unification* as a proposed premise.

If a postulate leads to a contradiction, then it can be said to be disproved. Some will improperly claim my work is confronted by contradictions and is therefore invalid. However, those false claims of contradiction will be irrelevant to the proper application of logic for scientific thought. They will be based upon blind acceptance of 'anti-physical' abstractions and the erroneous interpretation of observations and experiments. Conversely, as I expose the contradictions underlying today's physics, please consider this as evidence against the status quo mantras of Modern Physics.

Heretofore we have been unable to understand Modern Physics within the context of traditional *Common Sense Physicality.* I believe we have been duped into accepting abstract ideas that are 'anti-physical'. I believe ideas such as 'instantaneous non-local causality', 'parallel universes', and 'extra dimensions' are unworthy indiscretions. These are primarily tenets of Quantum Mechanics, though Relativity and String Theory also become entangled with 'extra dimensions'. Somehow, we have let 'anti-physical' abstractions supplant our instinctive good sense of physical reality.

Given our known observations and our improper mathematical formulations, sound and proper logic compels us to accept the 'anti-physicalities' of today's physics. Conversely, traditional common sense leads us to question these 'anti-physicalities'. If the logic is correct and the conclusion is wrong, then one or more premise must be wrong. I contend that the mathematical presentations of physics are flawed in some very subtle ways. By correcting the mathematical formulations with slight changes, we can completely alter the results. We can still explain the observations properly attributed to today's physics, but we can also restructure Modern Physics to conform to a viable vision of *Common Sense Physicality.*

With my *Postulate of Physical Unification*, I thus draw a line in the sands of evolving scientific thought. Will the 'blowing winds' of peer review erase this line and doom this vision to oblivion? Or will this vision emerge to serve as the mold for recasting scientific truth and thus become an unexpected milestone in the history of science? This proposed scientific revelation is being presented for your personal consideration. I encourage you to read onward with cautious curiosity and to make your own judgement, and I hope you can share my judgement in the affirmative.

Chapter Article 1.3

Einstein's Two Energy Equations

In his miracle year of 1905, Einstein promoted two fundamental equations about energy, E=hf and $E=mc^2$. The equation E=hf is the energy equation of photons (the particles of light which have no rest-mass). The equation $E=mc^2$ is the energy equation of ordinary matter (the particles of ordinary matter that do have rest-mass). Every particle either has rest-mass or does not have rest-mass, but whichever case it may be, every particle has 'self-energy'. The existence of particle 'self-energy' is a necessary quality of particle existence, and the total quantity of energy is uniquely conserved. The 'Law of Energy Conservation' is the most basic law of physics and states that the total amount of energy in the universe is conserved. In 1905, Einstein provided a totally new basis for us to consider the elemental forms of energy.

Einstein applied the energy equation E=hf to the 'photoelectric effect', and for this revelation he eventually received the Noble Prize in 1921. This is the energy equation for photons, the particles of light also known as the quanta of light. The 'E' represents energy, the 'h' is Planck's Constant, and the 'f' represents the frequency of photons. Each photon independently possesses some amount of energy and a corresponding frequency.

Max Planck originated the equation E=hf in 1899 to describe the quantized emission process of light as the underlying cause for the intensity/frequency profile of blackbody radiation. Blackbody radiation is merely the intensity of the various colors of light emitted by any object. When an object is heated its blackbody radiation increases in all frequencies, but the frequency of peak intensity shifts to a higher frequency.

Planck introduced the equation E=hf to describe quantized emission but not quantized transmission. Einstein revealed the equation E=hf describes light transmission as also being quantized. Light consists of traveling quantized bundles of energy, which we now call photons. Light thus paradoxically possesses both wave and particle characteristics in what is called 'wave-particle duality'. Since these particles of light never exist at rest, they possess no rest-mass.

Since light frequency corresponds to the wavelength and also the color of light, E=hf expresses the idea that photon energies vary with light color. Blue light photons possess higher frequency and thus higher energy than red light photons. Einstein's General Relativity revealed that gravity causes escaping light to appear redshifted, but the story of *Gravitational Redshift* has been misrepresented.

Modern Physics has failed to realize that gravity does not alter the free fall self-energy of photons. It is falsely assumed gravity reduces the energy of photons attempting to escape a gravitational field to thereby induce a redshift. For Einstein, gravitational time dilation induced a source within a gravitational field to emit photons of lesser redshifted frequency, and subsequent transmission did not alter that emitted frequency. I believe Einstein 'WAS RIGHT' about *Gravitational Redshift*, but his description was 'NOT QUITE' complete, because he ignored the issue and implications of photon energy. (He virtually ignored how gravity alters $E=mc^2$.)

In addition to applying E=hf to photons in 1905, Einstein developed $E=mc^2$, which has become the most recognized equation throughout the world today. This is Einstein's energy equation for rest-mass particles. The 'E' represents energy, the 'm' represents mass, and the 'c' represents the speed of light. With the equation $E=mc^2$, Einstein introduced the idea that energy and mass are somehow interrelated. Mass cannot only be converted into energy, but mass itself is a form of energy. This subsequently formed the basis for developing nuclear power and the atomic bomb.

When Einstein introduced $E=mc^2$, he was not thinking in terms of nuclear power or atomic bombs. He was contemplating the issue of Special Relativity and the 'Invariance of the Laws of Physics' for all observers in inertial motion. (Moving inertia observers could imagine themselves to be at rest and simply apply the Laws of Physics for a rest observer.) Einstein's first paper on Special Relativity emphasized the ideas of relative length contraction, relative time dilation, and the constancy of the measured speed of light. He understood that his initial presentation of Special Relativity was incomplete. He needed to also reframe the ideas of mass, force, momentum, and energy to be compatible with his vision of measurement invariance for moving observers. So, Einstein's second paper on Special Relativity in 1905 introduced $E=mc^2$ to complete his vision of relative measurement invariance.

The ingenious characteristic of Einstein's thinking about Relativity was that he sought brief statements of broad scope.

The brevity of the statements made them seem so elegant when contrasted with the boldness of their broad scope. Elegant simplicity, broad scope, bold generalization, and an aura of mystery are what made Einstein's Relativity seem so profound. To develop $E=mc^2$ from his initial vision of Special Relativity seemed to border on reckless speculation to many physicists of his day. Nevertheless, he faithfully followed the implications of his initial tenet for the 'Invariance of the Laws of Physics'. This led Einstein to develop his famous energy equation $E=mc^2$.

Prior to 1905, it was known that fields surrounding a particle possess energy, and thus a particle was understood to possess some self-energy. One proposal depicted the self-energy as $E=(4/3)mc^2$. While this seems quite close to the final version, the significance attributed to $E=mc^2$ by Einstein implied much greater fundamental importance. $E=mc^2$ is the self-energy of ordinary matter, and Einstein revealed that motion somehow induced a change of mass to manifest a change of this self-energy. This change of self-energy turned out to be the hidden form of kinetic energy, which is the classical energy of motion. To the extent that motion changed mass, motion must change the self-energy of $E=mc^2$ for those particles that possessed a rest-mass.

I suggest that, ultimately, we need to consider both the exterior field energy and the internal field energy of a particle. We shall want to consider how a field can form a self-contained bundle of energy. Pursuing that fundamental insight is the key to unraveling the mystery of *Physical Unification*. That is the topic of a later article in this chapter about *Cosmutons*.

For ordinary matter, Einstein introduced the energy equation of $E=mc^2$ with a special factor 'γ', (the lower case Greek letter gamma), to reflect the relative distorting effect of relative motion.

$$\boldsymbol{E} = \boldsymbol{m_o c^2} \gamma$$

The term γ varies with relative motion from one to infinity and is defined by the following expression:

$$\gamma = \boldsymbol{1 / \sqrt{1 - v^2 / c^2}}$$

The constant '***c***' is the speed of light, and the variable '***v***' is the relative velocity of the object being observed.

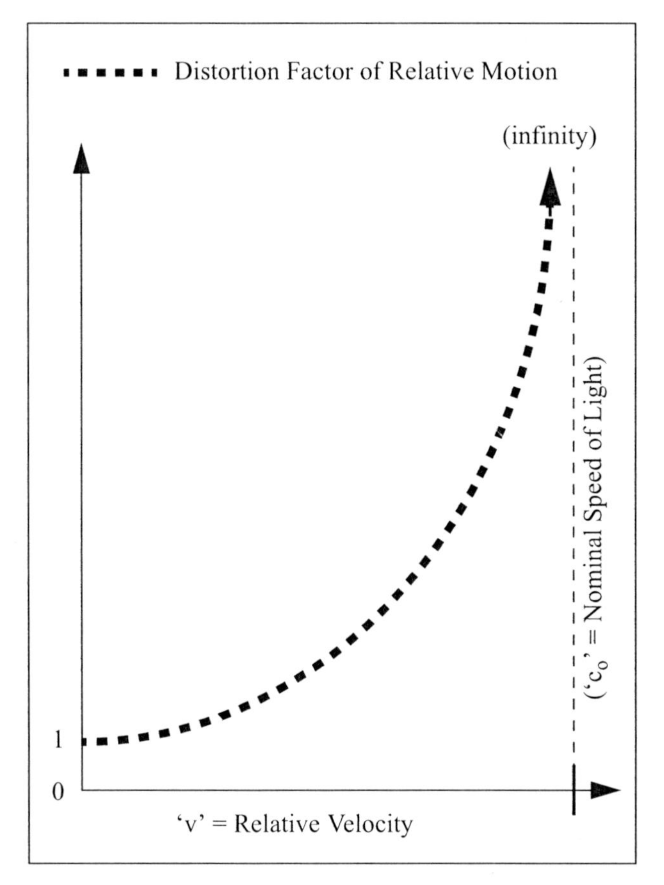

Figure-1: Relative Motion Distortion Factor

When an object is at relative rest the value of 'γ' is one. If an object could get up to the speed of light, (which it cannot), the value of 'γ' would become infinite. At the speed of light 'γ' would become infinite (an impossible physicality), which underlies why objects with rest mass can never reach the speed of light. Their self-energy would become infinite and that is impossible. *Figure-1* presents 'γ', which we can call 'Einstein's Gamma' or 'The Distortion Factor of Relative Motion'.

The term 'm_o' represents the mass at rest, or rest-mass of an object, and was considered to be a constant. However, when an object is in relative motion, it appears that its relative mass is changed. The relative mass is altered by the relative motion factor γ. Einstein's relative mass of motion 'm' was simply equal to '$m_o\gamma$'.

$$m = m_o\gamma$$

When a mass is at relative rest, the relative motion factor is one, so that the relative energy at rest 'E_o' is then equal to 'm_oc^2'.

$$E_o = m_oc^2$$

As an object begins its relative motion, the relative motion factor increases; and likewise, the relative mass and the relative energy increase. If the relative velocity could reach the speed of light, the relative motion factor would increase to infinity, and the relative mass and relative energy would also become infinite. However, the relative velocity can never reach the speed of light, so the relative mass and relative energy always remain finite.

Some of today's physicists confuse Einstein's version of $E=mc^2$ by wanting to ignore the idea of relative mass and only think in terms of relative energy. When Einstein proposed mass-energy equivalence with relative mass increasing with relative motion in conjunction with spacetime distortions, he established a sufficient basis for upgrading classical mechanics to relativistic mechanics. To ignore the variance of relative mass is to ignore how Einstein sought to incorporate Newton's famous force equation, F=ma, into the Special Theory of Relativity. Rather than promoting Einstein's updating of classical mechanics, these Modern Physicists appear to repudiate classical mechanics. This charade of ignoring the change of mass has been an unhelpful diversion, so it will itself be herein ignored, because it leads away from the actual physicality we seek to discover.

Einstein failed to proclaim the constancy of the energies E=hf and $E=mc^2$ for particles free-falling in a gravitational field. He also failed to properly account for the gravitational effects upon a rest-mass particle at rest in a gravitational field. Even more fundamental, Einstein failed to describe the effects of absolute motion upon rest-mass particles and to reconcile absolute motion with relative motion. These miscues within Relativity, his acclaimed area of expertise, throttled Einstein's other quests. Einstein was thus weakened in his questioning of the 'anti-physicalities' of Quantum Mechanics, and he was impaired in seeking his cherished dream of a 'Unified Field Theory'.

The greatest mistake attributed to Special Relativity involves the misinterpretation that Relativity proved a medium of space does not exist. While Einstein initially contributed to that mistake, he tried to recover from it after he developed General Relativity. In 1920, Einstein prepared a written lecture, which he delivered at the University of Leyden. This Leyden lecture has been published in its entirety in a book entitled "*SIDELIGHTS ON RELATIVITY*". In this lecture, Einstein unequivocally stated that gravitation without a medium of space was "*unthinkable*". Einstein eventually came to call this medium of space the 'Total Field'. The 'Total Field' was to be the basis of his 'Unified Field Theory'. Einstein's Leyden lecture is conveniently ignored, because it stands in stark contrast to the current interpretations improperly attributed to Einstein. Einstein's 'aether' and 'quantum reality' are excused as forgivable errors of a genius.

I believe (contrary to common belief) that Einstein's goal and his vision for *Unified Physicality* were essentially correct. Einstein recognized that force fields possessed energy but particles also possessed energy. Could it be that the E=hf energy of photons and the $E=mc^2$ energy for rest-mass particles are both just two particular manifestations of field energy? Einstein specifically thought that force field energies of a continuum did indeed form the various particles and that is why he called his vision of unification 'The Unified Field Theory'.

I believe Einstein's vision of 'The Unified Field Theory' was essentially correct. However, Einstein allowed the abstraction of an 'Abstract Field' to cloud his good sense. Einstein realized he needed a medium of space for gravity, but he failed to realize that he needed the properties of a 'physical' medium of space. That is why I emphatically seek a *Unified Physical Field Theory* and emphasize the physicality of a *Physical Field.*

Chapter Article 1.4

Gratke's Two Energy Equations

With Einstein's apparent acquiescence, Modern Physics has misrepresented energy as depicted by the two famous energy equations, $E=hf$ and $E=mc^2$. Modern Physics ignores and/or misrepresents how gravity affects the energy of both rest-mass particles and non-rest-mass particles. Even more basic, today's physics ignores the subtle effects of absolute motion on rest-mass particles. (NOTE: I define absolute motion to be real physical motion with respect to the *Cosmic Continuum*, the physical medium of three-dimensional space.) By initially ignoring the physical medium of space, Einstein misled Modern Physicists to forsake a golden opportunity to unravel the hidden story underlying *Ultimate Unification.* My revolutionary vision explains how to properly extend the story of energy into the realms of absolute motion and gravitational fields to alter our understanding of Quantum Mechanics, Particle Formation, and Cosmology.

Cosmological redshift relates to the reducing of expected light frequencies induced by departing Doppler motion, distance, and the supposed expansion of the universe. *Gravitational Redshift* relates to the reducing of expected light frequencies induced by gravity. Modern Physics appears to dismiss Einstein's most elementary explanation that gravitational time dilation induces *Gravitational Redshift*, but I claim that Einstein 'WAS RIGHT'. Einstein theorized (and experiments subsequently confirmed) that gravity slows all processes to manifest gravitational time dilation. However, for other physicists, Einstein was evidently 'NOT QUITE' convincing enough about this cause for *Gravitational Redshift.* These other physicists have been properly concerned about the energy of photons. How do the photons lose energy to become gravitationally redshifted? Modern Physics says the photons lose energy in the process of escaping gravity, so that photons 'supposedly' lose their energy in transmission. Modern Physics thus proclaims that photon frequencies must also be altered while in transmission through a gravitational field. However, this directly conflicts with Einstein's proclamation that time dilation at the source induces the entire frequency shift at the moment of photon creation and emission.

I contend that gravity does not alter the E=hf energy of photons in transit. To account for the altered energies of a photon escaping a gravitational field, the energy of the emitting source must also be diminished by the same exact factor that depicts gravitational time dilation. Thus gravity reduces the $E=mc^2$ rest-energy of ordinary matter. Einstein appeared to ignore this issue of energy, and Modern Physics appears to misunderstand this issue of energy. This failure to connect BOTH gravitational time dilation AND the gravitational reduction of rest-energy of ordinary matter is an enormous mistake that undercuts the integrity of General Relativity. The resultant fallacies lead to the unacceptable 'anti-physical dogmas' of General Relativity that I will explicitly debunk. By correcting our perspective of *Gravitational Redshift*, we can begin to reveal four major corrections to General Relativity:

- The distortions of mass and the speed of light by gravity combine to induce $E=mc^2$ to diminish its rest-energy, and this conversion of rest-energy constitutes the release of the 'potential energy of gravity'.
- The distortions of clocks, measuring rods, and the speed of light by gravity combine to produce an invariant localized measurement for the speed of light, even though the actual speed is reduced.
- The distortions of clocks, measuring rods, and the speed of light also induce a non-Euclidean spacetime geometry that does NOT curve into any 'extra dimension', as previously implied.
- 'Blackholes' are a fallacious conjuring induced by the incomplete and faulty formulation of E=hf and $E=mc^2$. In lieu of 'blackholes', very dense objects should now become known as *Greyholes*.

Had Einstein more fully pursued his natural instincts about *Gravitational Redshift*, he would have likely realized his omission about the effect of gravity upon $E=mc^2$. He also would have likely prevented others from improperly imagining that his curved spacetime required a fallacious 'extra dimension'. Einstein also thought that 'blackholes' were un-physical and therefore that they did not exist. Einstein had believed gravity progressively slowed light entering into a gravitational field, and this perception did not encompass the idea of a 'blackhole'.

It would have been difficult for Einstein to become very interested in this business about 'blackholes', because the term 'blackhole' was not even coined until after his death. 'Dark star' was the classical term used to describe a 'blackhole', and many simply doubted their existence. Unfortunately, Einstein failed to apply his explanation of *Gravitational Redshift* and the gravitational slowing of light to expose the fallacy of 'blackholes'. Had he done so, Einstein probably would have made greater progress toward his vision of a 'Unified Field Theory'.

With $E=mc^2$, Einstein revealed that relative motion somehow induced a change of perceived mass and relative energy. The amount by which the rest-energy was increased was to be the relativistic kinetic energy. Classical kinetic energy was considered to be the energy of classical relative motion. Einstein showed that classical kinetic energy was only a low speed approximation of the actual relativistic kinetic energy.

In contrast, I believe both absolute motion and gravity induce changes within $E=mc^2$ to manifest both 'absolute kinetic energy' and the 'potential energy of gravity'. I therefore introduce a revolutionary new version of the energy equation $E=mc^2$ by including two distinct distortion factors, 'γ_M' for absolute motion and 'γ_G' for gravity.

Einstein's Version: $$E = mc^2 = m_o c^2 \gamma$$

Gratke's Version: $$E = mc^2 = m_o c_o^2 \gamma_M^1 \gamma_G^{-1}$$

All three distortion factors, (Einstein's original 'γ' for relative motion, my 'γ_M' for absolute motion, and my 'γ_G' for gravity), range from the values of one to infinity. The exponent of plus one on 'γ_M' serves to emphasize an increase by one factor of absolute motion distortion. The exponent of minus one on 'γ_G' indicates a decrease by one factor of gravity distortion.

(NOTE: I employ a plus exponent on a gamma to mean multiply to increase, and a minus exponent to mean divide to decrease. I define my gammas specifically so that a plus exponent means bigger and a minus means smaller.)

The term 'γ_G' varies with the strength of the gravitational field, from one to infinity, and is defined by the following expression:

$$\gamma_G = e^{GM/rc_o^2}$$

The constant 'e' is the base of the natural log system and its value is a little less than three (2.7183...). The constant 'G' is Newton's Universal Constant of Gravitation. The constant 'c_o' is the nominal speed of light. The variable 'M' is the mass serving as the source of gravity, and the variable 'r' is the radial distance from that source.

I have determined that the term 'γ_G' has the particular form of an exponential because it satisfies two key requirements. It satisfies all direct observational evidence within our solar system (especially the perihelion precession of planets), and it precludes the physical absurdity of a 'blackhole horizon'. Thus, the extrapolation of observed effects in our solar system does not require us to conjure up the 'non-physical' idea of a 'blackhole'.

NOTE: In the past, some have prosed an exponential factor, but other errors of omission created the false impression that the exponential was inadequate.

The 'Inverse Distortion Factor of Gravity' can be a more convenient expression for historical comparisons:

$$\gamma_G^{-1} = e^{-(GM/rc_o^2)}$$

When an object is outside the influence of gravity the value of 'γ_G' is one. As an object moves very close to a would-be gravitational point-mass '**M**', the value of 'γ_G' becomes very large and approaches infinity. The 'Inverse Distortion Factor of Gravity' correspondingly goes from one down to zero, and *Figure-2* presents this inverse factor. This 'Inverse Distortion Factor of Gravity' is the factor by which clocks slow down, measuring rods are contracted, and the rest energy of $E=mc^2$ is reduced. *Physical Reality* requires that these quantities not become negative.

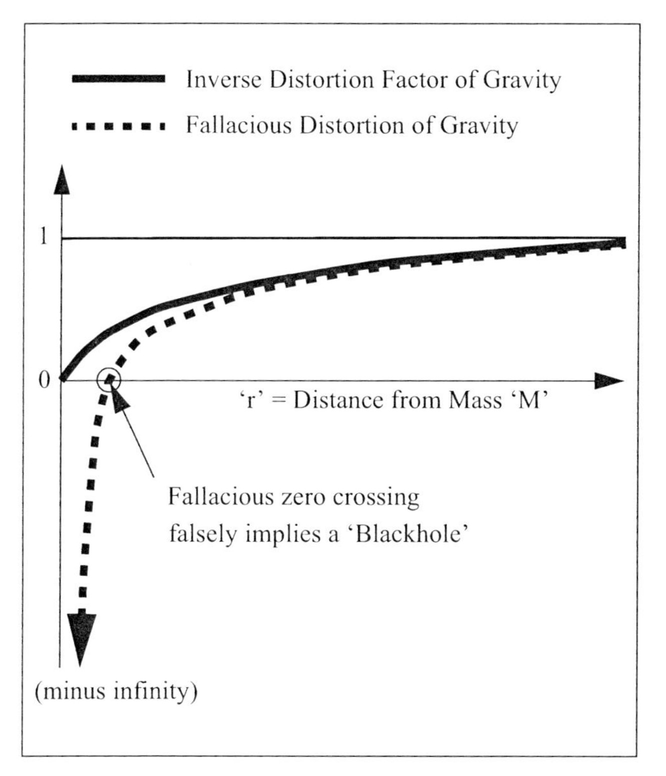

Figure-2: Inverse Distortion Factor of Gravity

A falling object gathers speed to increase its kinetic energy, but in falling, it loses gravitational potential energy. In free-fall the potential energy of gravity depicted via 'γ_G' with a negative exponent is being converted into the kinetic energy depicted via

'γ_M' with a positive exponent. The free-fall energy remains constant but is attributed differently. If something else were to absorb the falling energy so the falling object comes to rest, then 'γ_M' becomes one and the rest-energy '$\boldsymbol{E_o}$' is reduced by the amount indicated by 'γ_G' with the negative exponent.

$$\boldsymbol{E_o} = \boldsymbol{m_o c_o}^{\boldsymbol{2}} \gamma_G^{\boldsymbol{-1}}$$

Einstein's explanation that gravitational time dilation induced *Gravitational Redshift* was correct but incomplete. I proclaim gravity reduces the $E=mc^2$ rest-energy '$\boldsymbol{E_o}$' to induce a gravitationally redshifted proton to be energy diminished at the moment the photon is created and emitted. Time dilation inducing *Gravitational Redshift* should have implied to everyone that the $E=mc^2$ rest-energy of matter in a gravitational field is also reduced. Heretofore, no one has convincingly made this logical connection.

The mass '$\boldsymbol{m_o}$' should now be understood to be increased by both absolute motion and by gravity. Absolute motion increases the mass by one absolute motion distortion factor 'γ_M', and gravity increases the mass by a triple factor of the gravity distortion factor 'γ_G'. Instead of writing the gravity factor three separate times, we merely write the exponent three to so indicate. We can thus say gravity alters mass by the cube of the gravity distortion factor 'γ_G'.

$$\boldsymbol{m} = \boldsymbol{m_o} \gamma_M^{\boldsymbol{1}} \gamma_G^{\boldsymbol{3}}$$

(NOTE: Some have incorrectly thought that gravity might decrease mass to reduce the $E=mc^2$ rest-energy of matter. This does not work out for explaining the perihelion precession of planets. That the gravitational mass increases by a triple factor will be clearer once we clarify the gravitational slowing of light.)

While mass increases by a triple factor, light slows by a double factor of gravity, where '$\boldsymbol{c_o}$' is the nominal speed of light.

$$\boldsymbol{c} = \boldsymbol{c_o} \gamma_G^{\boldsymbol{-2}}$$

This double gravitational factor slowing of light conforms to the amount of slowing explicitly envisioned by Einstein. This slowing of light is also fully compatible with the confirmed observations that the path of star light bends and radar echo signals are delayed when they pass close to the sun.

Einstein's "Distortion Factor of Relative Motion' employs the value of the speed of light as a constant. However, gravity alters the speed of light, so it is no longer a constant. The speed of light is a variable. The reduced speed of light affects 'γ_M', which I now call 'The Distortion Factor of Absolute Motion'. As shown in *Figure-3*, as you descend deeper into a strong gravitational field the vertical limit on the graph shifts to the left.

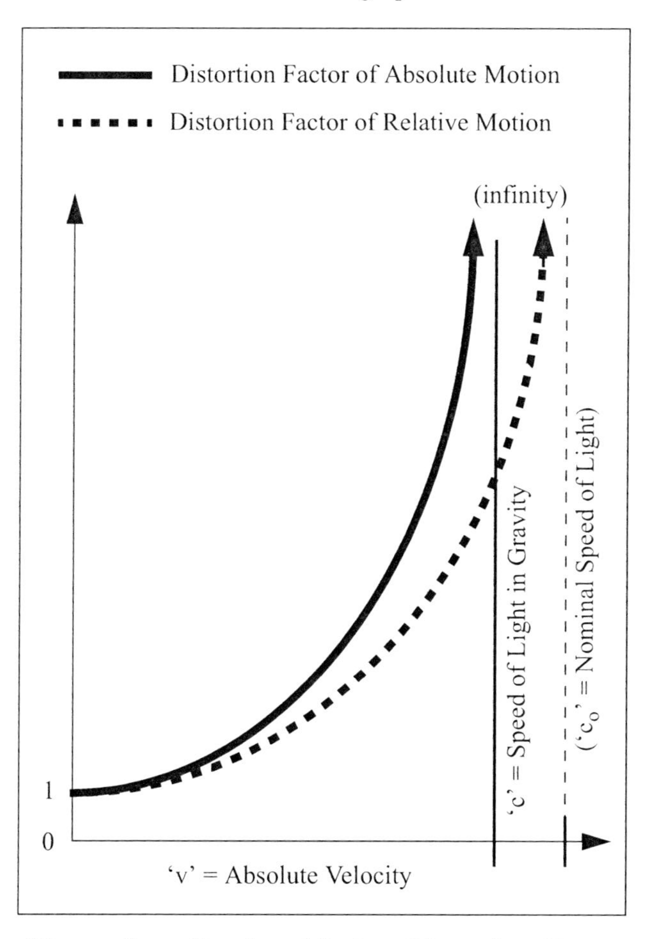

Figure-3: Absolute Motion Distortion Factor

When the increase of gravitational mass is combined with the decrease of the speed of light within the equation $E=mc^2$,

We see something Einstein never saw!

$$E = mc^2 = \{m_o \gamma_M^{\,1} \gamma_G^{\,3}\}\{c_o \gamma_G^{\,-2}\}^2$$

When we combine the triple increase factor of gravitational mass with the twice-doubled gravitational factor slowing light, we see the single decrease factor of gravitational potential energy.

$$\gamma_G^{(3-2\times 2)} = \gamma_G^{-1}$$

(NOTE: The triple increase factor of gravitational mass was deduced, knowing that gravitational potential energy must decrease by a single factor when light is decreased by a double factor.)

My revelation is that we can see the factors for both 'absolute kinetic energy' and 'gravitational potential energy' within the single energy equation $E=mc^2$!

$$E = mc^2 = m_o c_o^{\,2} \gamma_M^{\,1} \gamma_G^{\,-1}$$

The total energy for a rest-mass particle free-falling in gravity remains constant. The mathematical product of the kinetic energy factor times the gravitational potential energy factor appearing within $E=mc^2$ remains constant for free-falling particles. Einstein and virtually all Modern Physicists incorrectly portray the potential energy of gravity as a separate expression of energy. This separate expression for gravitational energy is to be added to the rest-energy and kinetic energy depicted within Einstein's simpler version of $E=mc^2$. For them, Einstein's $E=mc^2$ had no direct connection with gravity, and that has been a monumental blunder!

When I presented this revelation about $E=mc^2$ within a gravitational field to a few physicists at the April 2005 APS Meeting,

I was surprised at their indifference. To seriously consider my proposition, they would have had to alter just too many of their well established perceptions. They asked, "*What is the field equation?*" The 'field equation' is a 4-dimensional 4th order tensor with 256 elements (4x4x4x4). It should show conservation of energy. I could present such a field tensor, but the complexity does not help us to better understand General Relativity. In fact, it seems to be a hindrance. My version of $E=mc^2$ is a superior statement of energy conservation. In the past, the mathematical complexity of the tensor has tended to mask the actual *Physical Reality.*

Such is the problem with sophisticated mathematics. The mathematical formulations tend to be a translation of a physical perception, but not a perfect translation. Some (if not much) of the physicality is lost in the translation. The prior 'field equation' conserves energy, but its combined rest energy and potential energy becomes negative, which is a physical absurdity. In a similar way, the 'spacetime metrics' and 'light cones' depict the constancy of the locally measured speed of light; but, in the translation, improperly promote the physical absurdity of 'extra-dimensionality'. While the mathematics is quite sound, perception of the underlying physicality is distorted. Such distorting mathematical translations beget a 'sophistry of anti-physicality'.

While Einstein understood that gravity required a medium of space to facilitate General Relativity, he failed to publicly articulate how such a medium related to Special Relativity. This appeared to leave standing his previous renunciation of a medium for Special Relativity. On page 147 (of the 15th and final edition from 1952) in his book "*RELATIVITY - The Special and the General Theory*", Einstein added the acknowledgement that the Michelson-Morley experiment did not contradict a physical medium of space. Nevertheless, Einstein allowed himself to dwell upon the 'Field' without ascribing any physicality to it. The 'Field' is a mathematical construct with no physicality per se; and, left to the mathematicians, it becomes a mathematical description of nothing. I specifically claim the medium of space is a real and physical entity, which I call the *Cosmic Continuum.*

I claim the *Physical Reality* is that objects become distorted by their absolute motion as well as by their presence within a gravitational field within the *Cosmic Continuum*. These *Absolute Distortions* tend to maintain the 'local measurement invariance' postulated by Einstein's Relativity. Whereas Einstein's Relativity tends to be viewed as rather mystical, this new model of *Absolute Distortion* exudes a sense of underlying *Objective Physicality*. Without the 'reference frame' of the medium, objects do not 'know' they are to become distorted. Thus, that Relativity is confirmed by observation serves to confirm that a physical medium does indeed exist.

Chapter Article 1.5

Unifying Newton's and Einstein's Equations

Newton and Einstein are the two most famous physicists of all time. In 1687, Newton introduced two revolutionary force equations, and in 1905 Einstein introduced his revolutionary energy equation.

Newton's Equation of Motion: $$F = ma$$

Newton's Law of Gravity: $$F_g = GMm / r^2$$

Einstein's Energy Equation: $$E = m_o c^2 \gamma$$

By 'Perfecting E=mc^2', I am able to show how all three of these famous equations can be derived from my version of E=mc^2.

Gratke's Energy Equation: $$E = m_o c_o^{\,2} \gamma_M^{\,1} \gamma_G^{\,-1}$$

The mathematics is more than I wish to show here, but I can describe the process in words. The rate of change of energy with distance is force. In the language of mathematics, force equals the mathematical gradient of energy. When we take the mathematical gradient of my version of E=mc^2, we can derive both of

Newton's famous equations as first order approximations. When we apply my version of $E=mc^2$ to a region absent of gravity, where my distortion factor of gravity 'γ_G' is one, we obtain the approximation of Einstein's version of the energy equation $E=mc^2$. The subtly of Einstein's $E=mc^2$ (with a relative motion distortion) is realized as a transformation of my version of $E=mc^2$ (with an absolute motion distortion).

The prior failure to connect the relative motion factor with an absolute motion distortion factor has been the unintended consequence of the mathematical elegance of Special Relativity. The elegance of Einstein's mathematical simplicity for Special Relativity has enticed us like a 'pied piper's melody'. The elegance causes us to ignore and misunderstand the *Objective Physicality* that underlies Relativity and all natural phenomena. Some will want to continue chanting the mantra of the relative distortion factor, and with myopic indifference, ignore the underlying *Objective Physicality* of our universe. I say please, no more obfuscation!

The null result of the Michelson-Morley experiment of 1887 is often cited today as the reason for rejecting a physical medium of space. In the 1890s Fitzgerald and Lorentz independently suggested that objects were distorted by their absolute motion to produce that null result. In 1905, Einstein presented his Special Theory of Relativity wherein absolute motion could never be measured, and therefore he claimed we did not need a medium of space. Einstein led us astray with that proclamation. Later, he more carefully said we can describe the observations of moving observers as if they are rest, so we do not need to consider a medium of space to mathematically describe the 'mathematical relationships between observables'. This caveat about 'observables' is a subtlety of language and philosophy that has undercut the perception of *Physical Reality*.

There are two facets to the phrase 'mathematically described observables', but today's physics only wants to acknowledge one of those facets. The facet of mathematical description is only half of the story, and it is the smaller half. The bigger story lies in the underlying physicality that produces the physical relationships that justify the mathematical description. Without a sense of the underlying physicality, we are left in a state of confusion that breads auras of mystery and mysticism.

Einstein backed away from his 1905 careless rejection of a medium of space, but history seems to ignore this fact. Those writing the history have been indoctrinated to believe that a medium of space does not exist. Thus the historians simply ignore Einstein's subsequent acceptance of a medium of space. The medium of space has numerous names. In classical times it was called the 'aether'. In the parlance of Special Relativity, it can be called 'The True Rest Frame', but Special Relativity is typically interpreted so that such a frame does not exist. Einstein called the medium of space the 'Total Field'. I call the physical medium of space the *Cosmic Continuum*.

In 1920, Einstein prepared a written lecture entitled "*Aether and the Theory of Relativity*", which he delivered at the University of Leyden. This Leyden lecture has been published in its entirety in a book titled "*SIDELIGHTS ON RELATIVITY*". In this lecture, Einstein unequivocally stated,

"According to the General Theory of Relativity, space without aether is unthinkable."

Einstein eventually came to call this medium of space the 'Total Field'. The 'Total Field' was to be the basis of his 'Unified Field Theory'.

In 1916, Einstein wrote a book titled "*RELATIVITY - The Special and the General Theory*". The book has had a long and successful run. I use it as one of the required texts in my college course "*Relativity and Cosmology*". For the 15th and final edition of the book in 1952, Einstein added Appendix 5 "*Relativity and the Problem of Space*". Nearing the end of his most remarkable life, Einstein offered one last embellishment and clarification. In that Appendix 5, Einstein declares,

"Concerning the Michelson-Morley experiment, Lorentz showed that the result . . . does not contradict the theory of an aether at rest."

In this Appendix 5, Einstein is still dancing around the issue of space and time, in a way that remains unsatisfying to me. He is trying to blend the elegance of 'mathematically described observables' with a philosophical perspective of physicality. Maybe something is lost in the translation from German to English. However, his mathematical elegance still seems to trump *Physical Reality*. I want to turn the situation around. I want to describe *Physical Reality* so that it can be supported by mathematical elegance. *Physical Reality* should be the revered master, and mathematical elegance should be the faithful servant.

Any one of Einstein's inertial frame of reference could conceivably be 'The True Rest Frame'. Since it is believed we can never identify 'The True Rest Frame', the philosophy of 'detached observationalism' wants to proclaim such a frame of reference is nonexistent. Thus, this philosophy ultimately dismisses the idea of an absolute motion as well as the distortions induced by absolute motion. Such rejection is not justified by mathematics. It is not justified by observation. It is not justified by *Objective Common Sense Physicality*. However, it is justified by a deliberate philosophy of abstractionalism, which I call the 'sophistry of anti-physicality'. Please, no more denial of *Objective Physicality*!

When objects move through the *Cosmic Continuum* of absolute space, such objects become distorted. The absolute motion distortion factor 'γ_M' is the factor that describes the various distortions induced by absolute motion. Objects are also distorted when they are positioned within a gravitational field, where the *Cosmic Continuum* itself becomes distorted. The gravity distortion factor 'γ_G' is the factor that describes the various distortions induced by gravity. These two distortion factors form the basis for 'Perfecting E=mc^2' and unify the basis of Newton's and Einstein's equations.

As physics evolved between the times of Newton and Einstein, the idea of energy came into use. Newton's force equations had their subsequent counterparts as classical energy equations. These subsequent energy equations can also be derived from my version of E=mc^2. When we take the first order approximations for the algebraic expansion of my version of E=mc^2 we obtain the classical expressions for kinetic and gravitational potential energy, as well as Einstein's expression of rest-energy.

First order expansion of Gratke's E=mc²:

$$E = m_o c_o^{\,2} \gamma_M^{\,1} \gamma_G^{\,-1} \Rightarrow m_o c^2 - \frac{GMm_o}{r} + \frac{m_o v^2}{2}$$

Einstein's Rest-Energy: $E_o = m_o c^2$

Potential Energy of Gravity: $PE_g = -GMm_o / r$

Kinetic Energy of Motion: $KE = m_o v^2 / 2$

When we take the second order approximations for the algebraic expansion of my version of energy, mass and angular momentum, we obtain a sufficiently accurate basis for calculating the perihelion precession of planets. This perihelion precession calculation is merely a small correction to the classical calculation of orbits (though the calculation itself is tedious).

When we derive both of Newton's equations from my version of $E=mc^2$ we show that the 'inertial mass' in Newton's first equation is identical to the 'gravitational mass' in Newton's second equation. Similarly, we show the 'inertial mass' in the kinetic energy equation is identical to the 'gravitational mass' in the gravitational potential energy equation. The exact equality of the two masses has long been a point of confusion and inquiry, but now we can see that both masses come from a single source term within $E=mc^2$.

Modern Physics is often depicted as having overturned Classical Physics, but this is MISLEADING and WRONG. We did not overturn, but we did refine and extend Classical Physics. I propose an '*Era of Unified Physics*', wherein both Classical Physics and Modern Physics are incorporated as previous approximations. I seek to unify the prior visions of physics under the more encompassing umbrella of *Unified Physicality*.

Chapter Article 1.6

$E=mc^2$ Now Alters General Relativity

Once we recognize that gravity increases mass and slows light to decrease the rest-energy of $E=mc^2$, we are then in a position to recognize other new insights. These insights include:

- The positiveness of energy.
- The unity of inertial and gravitational mass.
- The *Physicality* of perihelion precession for planets.
- The omni-directional contraction of lengths.
- The non-Euclidean character of space.
- The Un-Curved character of Spacetime.
- The *Physicality* of Three-Dimensional Space.
- The *Cosmic Continuum* (the physical medium of space).
- The *Compression Gradient* of Gravity.
- The *Gravio-Magnetic Force.*
- The fallacy of dark matter.
- The *Gravio-Reaction Force.*
- The positive distortion factor of gravity.
- The fallacy of 'blackholes'.
- *Greyholes* (which are to replace 'blackholes').
- *Greyhole-Quasars* (to replace primordial quasars).
- *Greyhole-Galaxy Evolution* (growth of galaxy clusters).

Positive Energy: The standard models of energy include the idea of negative energy, but I disagree with negative energy. In particular, I believe the depiction of the potential energy of gravity as a negative energy is wrong! The nominal potential energy of gravity is a reduction of the rest-energy depicted by $E=mc^2$ (mass increases but the speed of light decreases). In the sense that this is a reduction, it can be considered negative. However, the rest-energy cannot become a negative quantity. Dividing by the positive gravity distortion factor 'γ_G', which goes from positive one to positive infinity, we cannot diminish the rest-energy below zero. Total particle energy thus always remains positive. Einstein warned that his formulations of General Relativity were only approximate, so he would not have been drawn into accepting negative total energy. Nevertheless, Modern Physicists blindly extrapolate Einstein's approximations to promote the fallacious concept of negative energy.

Mass Unity: For centuries, physicists have marveled at the equivalence of the inertial mass of motion and gravitational mass. The two masses have appeared as separate entities first in Newton's force equations and later in Einstein's energy equations. The inertial mass appears within the expressions for inertial force and also kinetic energy, while the gravitational mass appears within the expressions for gravitational force and also gravitational potential energy. With my new energy expression for E=mc^2 and its subsequently derivable force expressions, all these references to mass have a single common source. That source is the term for rest-mass 'm_o' that appears just once in my expression of E=mc^2. With only one appearance of the rest-mass term, there is no reason to suggest any distinction between inertial and gravitational mass. The two masses are not just equivalent. They are identical. Inertial and gravitational mass are the very same physical quantity.

Perihelion Precession: Einstein produced a mathematical description for General Relativity that is very difficult to follow. Only a small percentage of people have sufficient mathematical skill to follow the description of basic effects such as the perihelion precession of the planet Mercury. Einstein belabored the subtle effects of curved spacetime upon the planet Mercury as it follows its accentuated elliptical orbit. As it rises high and slows down, it has one system of coordinates; and as it falls low and speeds up, it has another system of coordinates. Trying to follow the continuously distorting system of coordinates of the ever-changing curved spacetime requires great mathematical skill. Typically, such calculations are left to the most energetic of graduate students pursuing doctorate degrees in theoretical physics or mathematics. The average person has very little chance of understanding such mathematics. That confusion has served to blind virtually everyone about the actual underlying physicality. However, all anyone needs to know beyond classical high school physics is that mass changes due to absolute motion and gravity. Energy and angular momentum are conserved, but (as Einstein incompletely revealed) mass is not conserved. With sufficient skills in advanced high school mathematics, one can readily confirm perihelion precession. Basic calculus reveals that variable mass, variable speed of light, and variable rest-energy induces the observed perihelion precession of the planet Mercury.

Omnidirectional Length Contraction: Einstein leveraged his 'Principle of Equivalence' equating acceleration and gravity to deduce the gravitational slowing of clocks and the bending of star light by the sun. For him the bending of star light was induced by the slowing of light in a gravitational field. Once Einstein identified the gravitational slowing of clocks and the gravitational slowing of light, he felt obliged to proclaim that gravity also contracted lengths. Since this length contraction is the same in all directions, we can call it omni-directional gravitational length contraction. Einstein sought to maintain an invariant local measurement of the speed of light. Today's physicists have inconveniently lost sight of this very important perspective. I agree whole heartily with Einstein on this point. In a gravitational field, measuring rods at rest contract omni-directionally and clocks slow down. Using distorted measuring rods and clocks to measure the local speed of light then produces the nominal value for the speed of light. Gravity actually slows the speed of light by a double distortion factor of gravity, but the distortion of the measuring devices exactly compensates. The locally measured speed of light is thus a constant even thought the actual speed of light does vary. The significance of this subtly lies in its implication for 'non-Euclidean space' and a corrected interpretation of 'curved spacetime'.

Non-Euclidean Space: With the progressive distortion of lengths and time for objects in a gravitational field, Einstein realized that the measured mapping of space would produce a non-Euclidean geometry. He thus invoked the non-Euclidean geometry of Riemann, which had been previously developed for the surface geometry of curved surfaces. Einstein, and virtually all physicists, thereafter have cited the example of a curved surface to explain the non-Euclidean geometry of gravitationally distorted spacetime. Consequently, we regularly encounter the descriptive term 'curved spacetime'. Since two-dimensional surfaces curve into the third dimension, it has been logical to presume that three-dimensional space curves into an unspecified 'extra dimension'. However, this interpretation of 'extra-dimensionality' as the required basis for 'non-Euclidean Space' is terribly wrong! There is another cause for 'non-Euclidean Space'.

Un-Curved Spacetime: We can create a non-Euclidean surface geometry on a flat surface that does not curve into the third dimension. If the measuring rods are systematically distorted (progressively shortening) on a flat surface, we produce a non-Euclidean surface geometry. The geometry is distorted by variable scaling on a two-dimensional surface without curvature into a third dimension. In a three-dimensional gravitational field, the progressive distortion of measuring rods produces a non-Euclidean spacial geometry without curving into any extra fourth spacial dimension. Since gravity also slows all timing devices, gravity is said to 'curve spacetime'. However, the label 'curved spacetime' is quite misleading because it implies an extra dimension where none exists. Some noted physicists conjure up the fiction of wormholes that supposedly penetrate through this fallacious extra dimension.

Three-dimensional Space: The misperception associated with 'curved spacetime' has undercut the integrity of physicality. It suggests we can arbitrarily add 'extra-physical' constructions. If we can so easily conjure up an 'extra dimension', we then become easily tempted to conjure up 'blackholes' and other 'anti-physical' constructs. However, by limiting ourselves to only three actual dimensions of space, we can begin to understand how Modern Physics has been drawn into an unnatural way of thinking. By exposing the fallacy of a fictitious 'extra dimension' for 'curved spacetime', we psychologically prepare ourselves to expose more troubling 'anti-physicalities'. I thus seek to expose the much more egregious 'anti-physical dogmas' of 'blackholes', 'quantum mechanical mysticism', and the 'extra dimensions' of String Theory.

The Cosmic Continuum: The rejection of a physical medium of space has been the most monumental blunder induced by misinterpreting Special Relativity. The appearance of invariance for the laws of physics for observers in inertial motion is actually produced by the distortion of objects in absolute motion. For a skilled mathematician, it is a trivial conversion, but being biased by 'anti-physical' prejudice, such transformations are simply ignored. Einstein initially ignored this as well for Special Relativity, but for General Relativity, he realized he really did need a medium of space. As I point out the necessary distortion of ordinary objects by gravity, the distortion of the medium of space itself seems unavoidable. The existence of a physical medium of space is the most fundamental of all *Physical Realties*. Playing word games, abusing semantics, and employing 'sophistry' to avoid the physical medium of space has been an unforgivable avoidance of *Physical Reality.* Whereas Einstein used the somewhat vague term of 'Field' for the medium of space, I boldly proclaim the *Cosmic Continuum* to be the 'physical' medium of space.

The Compression Gradient of Gravity: By prematurely rejecting a physical medium of space, I believe Modern Physics has discarded the actual underlying physicality of gravitational dynamics. When we acknowledge a physical medium of space, we can see how a compression density gradient of the *Cosmic Continuum* can form the basic force of gravity. This is comparable to Newton's forgotten vision. This is also in keeping with Einstein's statement that, gravity without aether is "*unthinkable*", and his idea that, distortion of the field communicates the effects of gravity.

The Gravio-Magnetic Force: A changing gravitational field produces a *Gravio-Magnetic Force.* Even though this force is not truly magnetic, its generation so closely parallels how a changing electric field produces a magnetic force, that we misapply the term magnetic. This term is not my invention. The *Gravio-Magnetic Force* has yet to be directly observed, but it is a subtlety lurking within Einstein's General Relativity. It is the subtlety that justifies orbital calculations based upon current planet positions with no apparent delay of the gravitational signals. (The *Full-Book Edition* will better explain this point.)

No Dark Matter: The idea of dark matter has a reasonable basis. The proposed presence of dark matter would explain the anomalous orbital speeds of stars at the outskirts of galaxies. These stars are moving too quickly given the amount of mass we perceive to exist within the visible stars of the galaxy. To explain the quickened orbital speeds, today's physicists have proposed the existence of extra matter which we cannot see. Since we cannot see this matter, it is called dark matter. However, I propose another solution, namely, that gravity itself is altered from what it has been perceived to be. I contend that most, if not all, dark matter simply does not exist.

The Gravio-Reaction Force: I now add the caveat that the continuum's effect of gravity is a two way street. Newton's Third Law of Mechanics (the Law of Reaction) states that for every force there is an equal but opposite force. Consequently, if the continuum exerts a gravitational force on an object, the object exerts a 'reaction force' on the continuum. Since most of the mass of a galaxy is clearly in orbit, the mass, which creates gravity, also creates a reaction to alter the nominal gravity. I believe the *Gravio-Reaction Force* can account for the anomalous orbital speeds within galaxies without recourse to dark matter.

The Positive Distortion Factor of Gravity: The common sense ideas of positive energy, positive clock intervals, a positive speed of light, positive lengths, and positive mass imply a positive distortion factor of gravity. The detailed expression for γ_G cannot become negative. Just as self-energy should always remain positive, clocks should always run forward and not go backward in time. Time travel is simply impossible. Light should never slow to less than nothing. Likewise, the lengths of objects should never shrink to less than nothing. Mass also should never become negative. The expressions introduced by Einstein implied negative distortion in extremely strong gravitational fields. This did not bother Einstein, because he recognized and stated that his expressions were only approximate. However, Modern Physicists have improperly interpreted Einstein's potentially negative expressions to be exact. They foolishly accept the 'anti-physical' consequences implied by negative distortion in extremely strong gravitational fields.

No Blackholes: The idea of a 'blackhole horizon' is a direct consequence of blindly extrapolating Einstein's approximation for the distorting effects of gravity and by blindly assuming that photons escaping gravity lose energy. We should consider five factors related to the fallacious conjuring of 'blackholes'.

- The photons of light falling into a gravitational field slow down while escaping photons speed up. This is the exact opposite of the presumption that escaping photons slow down and fall back, which falsely delineates a 'blackhole horizon'.
- The $E=hf$ energy of a photon is not altered by its transmission through a gravitational field and cannot enable formation of a 'blackhole'. The energy of a gravitationally redshifted photon is diminished at the moment of its creation and emission.
- While the energy and the frequency of photons are not altered by transmission through a gravitational field, their wavelengths are altered by the changing speed of light.
- The $E=mc^2$ energy of the emitting source within a gravitational field is diminished. The rest-energy of the emitter is reduced by the exact amount needed to produce the observed *Gravitational Redshift* of energy for escaping photons.
- The distortion factor of gravity always remains positive. This directly contradicts the fallacy that the distortion factor of gravity becomes negative to create a 'blackhole horizon'.
- The idea that a 'blackhole' collapses to a point, known as a singularity of infinite energy density, is the antithesis of *Physical Reality*. 'Blackhole' singularities are foolish 'anti-physical' contrivances.

Greyholes: 'Blackhole' is a name given to a very dense object, to which mistaken properties are attributed. If what we have been calling a 'blackhole' is not black, what is the color of a very dense object? I say they are very dim and therefore grey, ranging from light grey to a very dark grey that is almost black.

Accordingly, we can say that very dense objects are *Greyholes*. Of course, the very term *Greyhole* is color neutral. The nominal emission colors are greatly redshifted as well as dimmed. So the nominal emission frequencies of visible light can be shifted into the near infrared, the far infrared, and then into the radio frequency bands.

Greyhole-Quasars: Quasar is an acronym representing a 'quasi-stellar radio source'. Quasars were discovered in the 1960s, and their radio emissions were identified as immensely redshifted light. The redshift was identified as being Cosmological Redshift. This redshift is 'supposed' to be interpreted as a function of distance, which then places quasars at the very edge of the universe. Halton Arp has overwhelming observational evidence to place quasars much closer to us and in near proximity to galaxies of much lesser redshift. I contend Quasars are just *Greyholes* that are emitting radio frequencies. The redshifts of quasars are not induced by Cosmological Redshift but by *Gravitational Redshift*. I believe this vision of *Greyhole-Quasars* vindicates the very much-maligned Halton Arp.

Greyhole-Galaxy Evolution: I believe a galaxy can feed material into its center to form a *Greyhole* and that *Greyholes* can form *Greyhole-Quasars*. I also believe that spinning *Greyhole-Quasars* form a super furnace that draws in and reformulates matter. The quasar formulation of matter could rival the big bang as an alternate source for producing the heavier elements. The intense internal pressures and rotational dynamics can then form spinning axial jets. These spinning *Greyhole-Jets* support the formation of spiraling electromagnetic fields that spew reformulated matter along the *Greyhole-Axis* to feed the formation of new galaxies. Dare I mention that this greatly enhanced significance of Electromagnetism happens to be in keeping with a prediction of the 'scientific heretic' Immanuel Velikovsky? Whether it does, or does not, is immaterial to the merit of electromagnetism, but it is rather curious.

Origins: I draw a line precluding me from making any bold claims about origins. I feel that making pronouncements about origins is just too speculative, and at best, these speculations are only educated guesses. I wish to reserve my guesses about origins for an article in the *Epilogue* of the *Full-Book Edition*. Nevertheless, let me just say here that I sense the current propositions of the big bang and dark energy are rather curious, somewhat dubious, and ripe for reevaluation.

As to the General Theory of Relativity, Einstein missed a unique opportunity. Unfortunately, Einstein failed to strongly follow up on his own sense of physicality and his beliefs that:

- His formulation of gravity was only approximate.
- Gravity progressively slows incoming light.
- An emitter's gravitational time dilation induces a redshift.
- 'Blackholes' are un-physical.
- Gravity requires a medium of space.

Had Einstein only pursued his natural instincts he surely would have made much greater progress toward his vision of a 'Unified Field Theory'. Unfortunately Einstein was trapped by mathematical formalisms and 'Extreme Quantum Weirdness'.

Chapter Article 1.7

$E=mc^2$ Now Suggests Relativity of Quanta

Einstein and virtually all Modern Physicists have failed to realize the physical effects of gravity upon the self-energy of $E=mc^2$. My revelation is that $E=mc^2$ in a gravitational field is altered for rest-mass particles at rest, but it is not altered for rest-mass particles in free-fall. Einstein essentially 'WAS RIGHT' about $E=mc^2$ for rest-mass particles in relative motion, but he was 'NOT QUITE' prepared to deal with the issue of how absolute motion and gravity affects $E=mc^2$. By unveiling the proper story about particle energy, I am revealing an essential clue for demystifying Quantum Mechanics.

The unrealized effects of gravity upon $E=mc^2$ and the false effects of gravity perceived for $E=hf$ have crippled the proper development of a viable unification theory. Once we overcome the misconceptions about these two energies, we enter into a new domain of thought. Clouds masking our thinking begin to dissipate, and we begin to perceive *Physical Reality* with new clarity.

All known observations confirm that gravity slows clocks and atomic processes and thereby induces a *Gravitational Redshift*. Sir Arthur Eddington, who first observed the bending of star light by the sun, explicitly stated that gravity created a variable index of refraction to progressively slow incoming light. All known observations are consistent with the idea that gravity slows light entering into a gravitational field. Most physicists today who seriously contemplate General Relativity balk at this proposition of gravity slowing light. Gravity slowing light conflicts with their 'anti-physical dogmas' of 'extra dimensions', 'curved spacetime' and 'blackholes'. We have allowed fallacious abstractions to be elevated to an ordained perception of reality that is false, and these fallacies are then employed to reject physical realities.

Einstein realized that increased mass due to motion would produce a perihelion precession for elliptically orbiting planets. The amount of precession suggested by Special Relativity alone is not enough to produce the planetary precession observed. I thus interpret this to mean that gravity itself also increases mass. Gravity both increases mass and reduces the speed of light, and these two factors combined reveal that gravity reduces the rest-energy of $E=mc^2$. That gravity increases mass three-fold and reduces rest-mass energy one-fold is literally confirmed by the observed perihelion precession of Mercury! And the observation of *Gravitational Redshift* should be interpreted as being induced by the gravitational reduction of the rest-energy of $E=mc^2$.

Given the proper story of how Relativity affects particle energy, we can now begin to entertain the topic of Quantum Mechanics. When gravity reduces rest-mass energy, it does so by reducing the inherent quantum state energy of rest-mass particle formation. These quantum states of reduced energy then induce photon emission energies to be diminished and thus redshifted. Amazingly, the corrected stories of energy for rest-mass and non-rest-mass particles now form the basis for unifying Relativity and Quantum Mechanics.

The E=hf energy of a photon is literally extracted from the loss of $E=mc^2$ energy of ordinary matter. For a rest-mass particle, a quantum state in one harmonic mode transitions to a quantum state in a lower harmonic mode. If the photon is emitted with reduced energy, then the energy of the emitter rest-mass particle's quantum states must also be reduced. This is the unifying connection between Relativity and Quantum Mechanics!

The changes of kinetic and potential energy within $E=mc^2$ for rest-mass particles are actually changes in its quantum state harmonic energy attributed to particle formation. The energies of given quanta smoothly evolve to realize an underlying continuity for quantum states responding to the physical medium of space. The essence for unifying Relativity and Quantum Mechanics is that absolute motion and gravity smoothly distort the energies of rest-mass particle quantum states.

Unraveling the story of *Gravitational Redshift* is a primary key for decoding the mystery of unification. The proper explanation for the *Gravitational Redshift* of photons involves three necessary factors:

- Gravity slows the timing of an emitter to produce the reduced photon frequency of gravity's redshift.
- Gravity reduces the $E=mc^2$ rest-energy of an emitter to produce the reduced photon energy of gravity's redshift.
- Gravity reduces an emitter's quantum state energies to produce reduced energies for quantum transitions. This reduced quantum transition energy reduces both the frequency and the energy of gravity's photonic redshift.

Why would absolute motion and gravity cause the energy states of quanta to vary smoothly? If we could understand how and why quanta smoothly vary, then we should begin to grasp the *Objective Physicality* underlying *Ultimate Unification*. Absolute motion and gravity alter the wave harmonics underlying particle formation. The particle harmonics thus experience different timing and this alters the particle's 'self-energy', including the energy of its quantum states.

When a trumpet player depresses one of the keys, he selects an alternate air chamber with a different sized resonant cavity.

This varies the harmonic frequency of the internal standing wave producing a very noticeable step in the altered pitch of sound emanating from the trumpet. When an elementary particle mutates to an altered quantum state of self-resonant frequency, it is like the trumpet's resonant cavity being switched to an alternate air chamber.

Conversely, when a trombone player moves the trombone's slide, he smoothly varies the size of the resonant air cavity. He thereby smoothly varies the harmonic frequency of the internal standing wave. While the slide is in motion, we hear a smoothly altering pitch for the sound stream emanating from the trombone. Absolute motion and gravity are like the trombone's slide in that they smoothly alter the frequency of the self-resonant wave that forms the harmonic quantum state of an elementary particle.

Thus quantum states can vary in two ways. Particle quantum states can jump when switching to an alternate mode of self-resonance, which corresponds to the sense normally attributed to quantum transitions. In addition, particle quantum states can also vary smoothly with a particle's absolute speed and with its changing position within a gravitational field. This is the secret underlying the unification of Relativity and Quantum Mechanics.

Elementary particles with rest-mass, such as electrons and protons, form as looping wave vortices. These particles are waves in self-resonance, and this is the type of underlying physicality suggested by String Theory (but now restricted to just three dimensions). The self-resonance changes when a particle changes its absolute speed or changes its position in a gravitational field. A gravitational field is just a progressive change in the density of the physical medium of space (a *Compression Gradient* of the *Cosmic Continuum*). The various changes to an elementary particle's wave vortex are the very changes needed to produce the effects of Einstein's Relativity. These various changes to an elementary particle's wave vortex are changes in the particle's underlying self-resonance, and this is the basic underlying physicality of Quantum Mechanics.

When a rest-mass E=mc^2 particle is in absolute motion or in a gravitational field, its change of self-resonance alters the particle's physical properties. The five main features of this change are 'Time Dilation', 'Length Contraction', 'Mass Enhancement', 'Energy Distortion', and 'Quantum State Distortion'. (NOTE: For the gamma distortion factors shown below, remember that plus exponents mean increase and minus exponents mean decrease.)

- ***Time Dilation:*** The time for an underlying wave to complete one loop around its vortex is increased. This induces 'time dilation' as time intervals increase and frequencies decrease.

  Time Interval: $T = T_o \gamma_M^{1} \gamma_G^{1}$

  Frequency: $f = f_o \gamma_M^{-1} \gamma_G^{-1}$

- ***Length Contraction:*** The size and shape of an underlying wave vortex is altered. Absolute motion does not alter lengths transverse to the direction of absolute motion. However, absolute motion does contract a vortex in the direction of motion, and this is called '*longitudinal motion length contraction*'. By contracting an object in only one direction the shape of a vortex is altered. Gravity contracts a vortex uniformly in all directions, which is called '*omni-directional gravitational length contraction'*.

  Transverse Lengths: $L_T = L_o \gamma_G^{-1}$

  Longitudinal Lengths: $L_L = L_o \gamma_M^{-1} \gamma_G^{-1}$

- ***Mass Enhancement:*** Mass is not the actual material, but is a property of how material objects resist changes to their motion. The ability of a particle vortex to resist changes of motion is alterable. Both absolute motion and gravitational fields increase the mass of the elementary particles that possess rest-mass.

  Mass: $m = m_o \gamma_M^{1} \gamma_G^{3}$

- ***Energy Distortion:*** Absolute motion increases the wave vortex self-energy (the energy of E=mc$^{2)}$, and this increase is the kinetic energy of motion. Gravity decreases the wave vortex self-energy (the rest-energy of $E=mc^2$), and this decrease releases the potential energy of gravity. When a particle that possesses rest-mass (an $E=mc^2$ particle) is free-falling in gravity, the change of rest-energy feeds the change in motional energy.

Energy of Matter: $$E = m_o c_o^{\,2} \gamma_M^{\,1} \gamma_G^{\,-1}$$

- ***Quantum State Distortion:*** Absolute motion increases the quantum state energies of $E=mc^2$ particles. Gravity reduces the quantum state energies of $E=mc^2$ particles at rest, exactly the same as the overall energy of ordinary matter.

In addition to these preceding five changes for rest-mass $E=mc^2$ particles, we must also note the behavior of the non-rest-mass E=hf photons.

- ***Conserving Energy and Frequency:*** Gravity does not alter the energy nor the frequency of photons in transit.
- ***Slowing the Speed of Light:*** Gravity does progressively slow the speed of incoming light.

Speed of Light: $$c = c_o \gamma_G^{\,-2}$$

- ***Shortened Wavelengths:*** Gravity also progressively shortens the wavelengths of incoming light.

Wavelength: $$\lambda = \lambda_o \gamma_G^{\,-2}$$

The 'Clarifying of E=hf' and the 'Perfecting of $E=mc^2$' thus suggests a unifying vision for Relativity, Quantum Mechanics, and String Theory. Even though distortions occur and the 'actual speed' of photons do vary, the locally 'measured speed' of light remains constant Absolute motion and gravity alter the quantum states of rest-mass particles. This unifying vision is based upon an *Objective Physicality* functioning within our *Common Sense* three-dimensional universe.

Chapter Article 1.8

Confronting 'Extreme Quantum Weirdness'

A pathway to *Unified Physicality* is opened by 'Clarifying E=hf' and 'Perfecting E=mc^2', but if 'Extreme Quantum Weirdness' cannot be overturned, *"there is no hope"*. If the 'anti-physicalities' of Quantum Mechanics are allowed to stand, the very ideas of *Objective Physicality* and *Unified Physicality* are doomed!

Relativity Theory has falsely implied one or two 'extra dimensions', and String Theory has falsely implied several 'extra dimensions'. However, these faulty misadventures of a few 'extra dimensions' appear to be minor indulgences compared to the infinite number of 'extra dimensions' erroneously implied by Quantum Theory! This idea of 'extra dimensions' for Quantum Mechanics is abetted in its 'Extreme Weirdness' by the 'anti-physical' ideas of 'distributed mystical existence' and 'instantaneous non-local causality'. The absurdness of 'Extreme Quantum Weirdness' is the very antithesis of *Common Sense Physicality*!

The formal difficulties of Quantum Mechanics began in 1926. While many physicists were trying to understand what Quantum Waves actually were, Max Born suggested they were 'non-physical probability waves'. Also, in that year, Werner Heisenberg proposed his 'Uncertainty Principle'. Einstein considered this uncertainty to be the consequence of limitations in the accuracy of measurement. However, Born and Heisenberg wanted to view the uncertainty as a lack of 'physical existence', because an object 'supposedly' spreads out into its 'non-physical probability wave'. Thus, by 1926, the issue became 'physical existence' versus 'non-physical existence'.

Many readers will find it hard to believe anyone could be serious about espousing the current outlandish propositions of Quantum Mechanics. The one example I like to cite is:

The moon ceases to exist when no one looks at it, and re-materializes should anyone sneak a peek at it.

Quantum Theory proclaims that objects normally exist as 'non-physical probability waves' that are distributed throughout an infinite number of 'parallel universes'. Should anyone then sneak a peek at an object, the very process of observation stimulates 'instantaneous non-local causality' to collapse the object into its familiar form. This is absolutely remarkable, and absolutely inane!

Consider another example. Imagine I have two identical marbles, one Black and one White. I enclose them in my cupped hand so no one can see the two marbles. Next, I shake my hand so that the concealed marbles roll around without anyone observing them. Without myself, or anyone else being able to see the marbles, I then slip one of the marbles into a space capsule and send it to the moon. So what is the color of the marble that remains in my hand? You would surmise that the remaining marble is either Black or White. Amazingly, Quantum Theory says both marbles are Gray, because they are in a state of 'quantum probability entanglement'.

Imagine further, that I then open my hand and reveal the marble happens is Black. You would say the Black marble remained here on Earth while the White marble was transported to the moon. Quantum Theory says both marbles remained Gray until the instant we observed the one marble on Earth. By observing that the marble on earth was Black, via 'instantaneous non-local causality' we supposedly caused the marble on the moon to switch from Gray to White. Most amazingly, Quantum Theory says the observation of one marble causes the 'entangled quantum probability wave' of the pair to collapse into their respective 'individual observed realties'.

The world and the universe are not large enough to accommodate both *Common Sense Physicality* and 'Extreme Quantum Weirdness'. They are so mutually exclusive, that no *Theory of Physical Unification* can appear to be viable until these weird tenets are renounced!

I proclaim:

Extreme Quantum Weirdness has become the greatest hoax in the history of science!

'Extreme Quantum Weirdness' has been a horrendous fraud foisted upon us by well-intentioned but seriously misguided theorists. The terms 'hoax' and 'fraud' usually imply a deliberate intent to deceive. However, I do not believe deception was intended. Nevertheless, the blatant disregard of *Common Sense Physicality* has been so irresponsible that the severity of the terms 'hoax' and 'fraud' seem justified.

Einstein objected to the very strange nature of Quantum Mechanics, but few listened to his warning. Quantum Mechanics might always seem somewhat strange, but those aspects of it, which promote 'Extreme Quantum Weirdness', are simply unworthy of science. Here are the main issues confronting the hoax of 'Extreme Quantum Weirdness':

- 'Actual Quantum Waves' are NOT 'anti-physical probability waves'. They are *Physical Waves* within a *Physical Field.* ('Actual Quantum Waves' exude some unique properties because they are formed by and interact with the wave substructure of elementary particles.)
- Objects DO NOT exist in a dissolved state distributed throughout our universe much less throughout an infinite number of 'parallel universes'.
- Objects DO NOT communicate via an 'instantaneous non-local causality'. Objects communicate via a '*Principle of Local Causality*'.
- Observations purporting to confirm the 'anti-physical dogmas' of Quantum Mechanics have simply been MISINTERPRETED.
- 'Extra dimensions' are FALSELY INVOKED in Relativity and String Theory, and even more so in Quantum Mechanics. Actual *Objective Physicality* dwells within a three-dimensional *Physical Field*, and this *Physical Reality* changes over time.
- Examples of 'Extreme Quantum Weirdness' should not be cited to provide IMPROPER anecdotal support for the

FALSE 'extra-physicality' and 'extra-dimensionality' of curved spacetime and String Theory.

By touting an aura of 'extra-physicality', the abstract aspects of 'Extreme Quantum Weirdness' promote a sense of mathematical mysticism that is 'anti-physical'. This is not science! The actual nature of Relativity, String Theory, and Quantum Mechanics is to be found within an underlying *Physical Reality* that can be understood from the point of view of *Common Sense Physicality*.

The $E=mc^2$ energy of rest-mass particles and the E=hf energy of photons can be exchanged in whole or in part with one another. These two forms of energy are just different states of dynamic distortion of the same underlying substance, the *Cosmic Continuum* of three-dimensional space. Distortions of the *Cosmic Continuum* constitute a *Physical Field* that can form the elementary force fields and the elementary particles. The physical variables of particle formations in three-dimensional space are the physical embodiment of quantized Strings, which have been falsely characterized as dwelling in 'extra dimensions'.

When it seemed impossible to explain reality within the constraints of three-dimensional space, Modern Physicists introduced 'extra dimensions'. Physicists introduced abstract representations that seemed to work mathematically. However, they ignored and abandoned the idea that their mathematics should represent a plausible *Physical Reality*!

Thus, there have been major misconceptions about Relativity, Quantum Mechanics, String Theory, Particle Formation, and Cosmology that need to be exposed and corrected. Many more details will be discussed later, but at this point of introduction, let us focus on the following main point. Modern Physicists are trapped by their mathematical formulations and fail to recognize the need to conform their physics to the realm of three-dimensional *Objective Physicality*. By 'Clarifying E=hf' and 'Perfecting $E=mc^2$', I claim to establish a new basis for formulating a viable theory of *Unified Physicality*. However, before we can move forward with my vision, we need to overcome the hoax of 'Extreme Quantum Weirdness'.

Chapter Article 1.9

Debunking 'Extreme Quantum Weirdness'

Einstein strongly objected to the strange character of Quantum Mechanics. In 1935, Einstein, Podolsky and Rosen proposed criteria to disprove the non-physical interpretation of the 'Heisenberg Uncertainty Principle'. Einstein was trying to justify rejection of the Heisenberg interpretation of Quantum Mechanics that foolishly dismisses 'physical persistence' and 'physical existence'. All such criteria are now known as EPR criteria, and all such experiments are now known as EPR experiments.

The 'Heisenberg Uncertainty Principle' states that no one can precisely observe a particle, such that no one can precisely measure both the position and velocity of a particle. Niels Bohr led the development of the 'Copenhagen Model of Quantum Mechanics', in which 'observational uncertainty' is interpreted to mean uncertainty of 'physical persistence' and 'physical existence'. The 'Copenhagen Gang' championed an interpretation of the 'Uncertainty Principle' that denied *Physical Reality*. Conversely, Einstein interpreted 'observational uncertainty' as merely a limitation in our ability to make accurate measurements. Einstein assumed particles maintained their 'physical persistence' and their 'physical existence'.

Einstein originally proposed an EPR test which would produce matched pairs of particles, with each particle of a pair going in opposite directions. This particular test would measure the position of one particle of a pair and the velocity of the other. Since the two particles of a pair are matched, we would then know the position and velocity of both particles in the pair. Doing so would thereby establish the 'physical persistence' and 'physical existence' of both particles in violation of the 'Heisenberg Uncertainty Principle'.

Einstein's originally proposed EPR experiment was virtually impossible to do, and it has never been performed. In addition, no EPR experiment of any kind had been performed during Einstein's lifetime. Bohr responded to Einstein's proposal by claiming that the matched pairs of particles were 'quantum entangled' and that their 'quantum probability waves' collapsed instantaneously whenever either particle was sensed. Einstein objected and claimed such a collapse required 'instantaneous non-local causality'. Bohr responded that is exactly the character of phenomena in the quantum realm of semi-existence. They thus agreed that Bohr's interpretation required 'instantaneous non-local quantum entanglement' and 'instantaneous Non-Local quantum causality'. However, they disagreed in the following way. Einstein thought that the obvious physical absurdity of Quantum Mechanics was sufficient to reject Bohr's interpretation. Bohr countered that Einstein's physical reality was only an abstraction of the mind and did not truly exist.

In the 1950s, David Bohm proposed an EPR test using spin correlations for particle pairs with matched spin. This would have been a more feasible experiment than Einstein's correlation of position and velocity. In the 1960s, John Bell proposed using polarization correlations for photon pairs with matched polarizations. Bell's photon correlation test seemed even more feasible than Bohm's spin correlation test. Bell developed his now famous '*Bell Inequality*' which notes the difference between the 'Quantum Entangled Model' and a 'Classical Spin Model'. (However, as we will discuss shortly, Bell should have employed a 'Classical Polarization Model'.) In the 1970s, initial photon correlation experiments provided promising results. In 1981, Allan Aspect performed the definitive experiment and reported his experiment proved that 'instantaneous non-local entanglement' is the correct model.

Over the last 25 years, numerous EPR experimenters have reported that 'instantaneous non-local entanglement' is correct and 'Local Objective Reality' is wrong. All these experimenters claim to confirm 'instantaneous non-local causality' and thus prove the universe to be non-physical and governed by 'non-local entanglement'. These results are cited to justify 'Extreme Quantum Weirdness'. Left unchallenged, these claims are sufficient to destroy any hope for realizing any theory of *Objective Physical Reality*. However, I reject these EPR claims. My startling proclamation is that these EPR results are a fantastic mistake!

The reported EPR results are WRONG! Let us now examine the critical details for some of these EPR tests. Given a pair of matched photons going in opposite directions, the EPR experiment requires the detection of the two photons after they each pass through a polarized filter. The issue is to determine the probability that both photons are detected as a function of the angle of orientation between the two polarizers.

The 'Quantum Model' of correlation predicts the probability of detecting both particles of a particle pair to vary as the cosine squared of the angle between the axes of two polarizer filters (an S-curved function). Bell led us to believe that the 'Local Reality Model' predicts the probability of particle pair detection to vary linearly with the angle between the axes of two polarizer filters (a straight-line function). Allan Aspect reported that the observed correlation did indeed vary as the cosine squared of the angle, so that 'supposedly' no model of 'Local Objective Reality' could be viable. This observation 'supposedly' proves that 'Extreme Quantum Weirdness' wins, and *Objective Physicality* loses. However, this is a major mistake, possibly the biggest blunder in the history of science!

Bell's model of 'Local Objective Reality' employed for comparison improperly depicts classical photons as being particles with classical spin, but, this is WRONG! Bell allowed himself to be deceived. He thought classical particle spin and classical wave polarization were the same, but that is simply NOT true!

Particles of light are polarized and they behave differently than particles with spin. For photons, the correct classical model of 'Local Objective Reality' produces a measurement correlation that also varies with the cosine squared of the angle between the axes of the two polarizers. This cosine square relationship of 'Local Objective Reality' has been known since the early 1800s, long before the development of Quantum Mechanics. This cosine square relationship of 'Local Objective Reality' is clearly presented in many of today's High School textbooks. Quantum Theorists have let mathematical elegance blind their good thinking. They equate the mathematical form of spin and polarization, which deludes them into thinking that spin and polarization are somehow equivalent. Spin and polarization are not equivalent.

Consider a simple test. One perfect polarized lens is laid on another with their axes of polarization aligned. For un-polarized light incident upon them, half passes through the lens pair. When the axis of one is rotated 45 degrees, a quarter of the light passes through the pair. When the axis of that one is rotated another 45 degrees (for a total of 90 degrees), no light passes through the pair. This seems like it might be a linear relationship, but it is NOT linear. If we plot the measured intensity of light transmitted by the pair of lenses versus the angle of axis separation, we would see the intensity varies as the square of the cosine of the angle, $I = I_0 \cos\phi^2$. The 'Local Objective Reality' of single photons passing through a pair of polarized lenses is NOT linear. You can easily demonstrate this model of 'Local Objective Reality' for yourself using two polarizing lenses and a light meter.

Now consider a different test. We generate two identically polarized photons going in opposite directions, and each photon is to pass through its own polarized lens. (We have two photons and two polarized lenses, and each photon passes through one lens). What is the probability both photons of a pair will pass? If the polarization axes of the lenses were rotated to be offset at 90 degrees, Bell thought no photon pair could pass through both lenses, but this is WRONG! If a photon pair is at 90 degrees to one lenses, it must be at zero degrees to the other. While one photon would surely pass through, the other would surely not pass, so the pair has no chance to pass through (just as Bell had thought). However, if a photon pair is at 45 degrees to both lenses, each photon has a 50% chance, so the pair has a 25% chance to pass through the lenses. This contradicts Bell's presumption that no photon pair can pass through both polarizers when they are set at 90 degrees axis separation. Over many photon pair orientations (given 90 degree separation of the polarization axes), the average for passage is 12.5% (1 of 8 pairs will pass). Bell's theoretical model of 'Local Objective Reality' for photons is WRONG!

Consider one more variation, where the two polarizers are aligned with zero degrees of separation. If the polarization of a photon pair is exactly aligned, the pair surely passes, and if aligned at 90 degrees, the pair surely will not pass. However, what will happen if the polarization of a photon pair is aligned at

45 degrees to the polarizers. The Quantum model would say one photon has a 50% chance of passing, and if one does pass it will turn the other photon so it also would pass. The pair would pass with a 50% chance, but I DISAGREE! With the pair at 45 degrees, each photon has a 50% chance of passing, so the pair thus has a 25% chance. Over many photon pair orientations (given zero degree separation of the polarization axes), the average for passage is 37.5% (3 of 8 pairs will pass). In contrast, the Quantum model predicts a 50% probability, so a clear distinction exists.

Figure-4 depicts how probability varies as a function of the angle of separation between the polarizers for the three models.

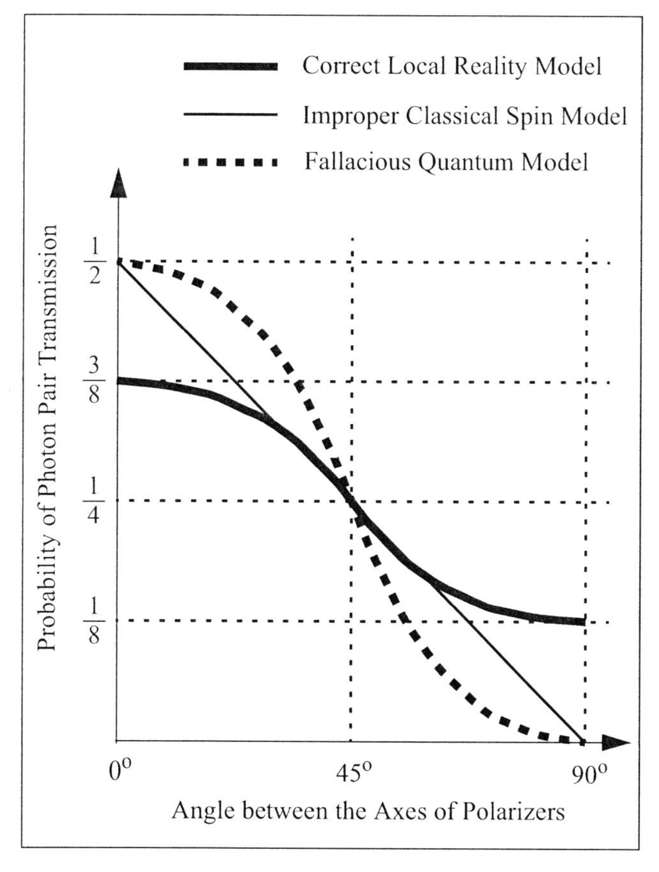

Figure-4: Photon Pair Correlation

The model of 'instantaneous quantum entanglement' would have us believe the unbelievable. If one photon should twist to get through its polarizer, it 'supposedly' twists the other photon to maintain entangled pair alignment. 'Supposedly', the polarized pair dwell in overlapping 'ghostly parallel universes'. 'Supposedly', when one photon was observed, it 'instantaneously collapsed' the other photon's 'quantum probability wave' into alignment within our universe. This of course is quite foolish!

Since the profiles of both models are now cosine squared, one might think that photon based EPR experiments could not distinguish 'instantaneous quantum entanglement' from 'Local Physical Reality'. However, the big surprise is that there is a clear distinction. As *Figure-4* depicts the three correlation profiles, you can readily see the difference for yourself.

This raises a VERY IMPORTANT question. Why did EPR experimenters NOT report seeing the profile of the Correct Local Reality Model? Some might falsely jump to the conclusion that this new model of 'Local Physical Reality' must be invalid, and that is why it was not reported. Here is the proper answer:

EPR experimenters did NOT report seeing the profile of the Correct Local Reality Model, because they were NOT looking for it?

When I addressed some physicists at the April 2005 APS Meeting with my revelation of Bell's error, I was amazed at their response. They wanted to know what I proposed as the proper connection between the two photons for a given split pair of photons. I tried to explain that there is NO connection whatsoever, but they could not accept that. They knew that there must be a connection, because EPR experiments have shown there is a connection. In trying to explain how EPR experiments are misinterpreted, the questioners kept falling back to the same old line. *"There has to be a connection, so what is the connection? How else could the EPR experiments be explained?"* They were trapped by their own preconceptions and could not realize that there is ***NO*** long distance connection - there is ***NO INSTANTANEOUS*** 'action-at-a-distance'.

Four factors combine to have blinded the EPR experimenters:

- Factor #1 - A Faulty Theoretical Model
- Factor #2 - Minimal Sensor Efficiency
- Factor #3 - False Data Pairs
- Factor #4 - Faulty Data Analysis

Factor #1 - A Faulty Theoretical Model: EPR experimenters were looking to distinguish a cosine profile from that of a straight-line profile. All they really concluded is that the profile they observed had the cosine shape, but they failed to distinguish beyond that distinction. Since the 'Improper Classical Spin Model' and the 'Fallacious Quantum Model' both go to zero at 90 degrees, they ignored the fact that actual test results were not zero at 90 degrees. Had they expected any of the proposed scenarios to have a non-zero value at 90 degrees they would NOT have rationalized away this very important feature in the data.

Factor #2 - Minimal Sensor Efficiency: The actual photon sensors employed are photomultiplier tubes, which may be only about 10% efficient. While one tube sees only 1 out of 10 photons, a pair of photomultiplier tubes can only see about 1 out of 100 photon pairs. Experimenters must thus generate many photon pairs to accumulate significant statistical data. EPR experimenters are thus 99% blinded to the actual data, and are forced to embellish the 1% of the actual data that they do have.

Factor #3 - False Data Pairs: What if you see two photons but they come from two different photon pairs? That would generate false data that would have to be rejected. Two pairs of photons do not need to be exactly simultaneous to be considered simultaneous for experimental purposes. Within a single sensor, two pulses could be close enough in time so that the two pulses appear as one larger blurred pulse. Given two separated sensors with two pulses, one in each sensor from a different photon pair, the pulses must merely be close enough in time to generate a false indication of a detected pair.

Factor #4 - Faulty Data Analysis: How does one distinguish a false indication of a detected photon pair? Such false data would have to be rejected. In fact, this false data does generate the same non-zero value at all angles including the angle at 90 degrees. Knowing about this possibility, EPR experimenters must have simply rationalized that all the non-zero data at 90 degrees must be produced in this manner. Accordingly, they simply subtracted all the pair counts at 90 degrees from the whole profile. EPR experimenters would then rescale the remaining profile so as to fit their remaining data into their preconceived profile of the 'Fallacious Quantum Model'. However, this faulty analysis can be overcome.

The EPR Correction: The percentage of multiple pulses sensed in just one sensor should equal the number of false photon pairs sensed by two sensors. Subtracting this observed probability of multiple detection by a single sensor from the profile of detected pairs will serve to reject false data pairs. However, it should still leave a significant number of detected pairs at the angle of 90 degrees. The net remaining probability at the zero angle should now be three times the net remaining probability at 90 degrees.

If the raw data of any previous EPR experiment actually detected, counted, and recorded multiple pulses in a single sensor, it will contain the information that would allow us to distinguish between all three models. I boldly predict such data will clearly and convincingly enable us to conclude:

- Bell's famous 'Inequality' for photon pair correlation is simply WRONG.
- Bohr's 'Standard Copenhagen Model of Quantum Mechanics' is fantastically WRONG!
- Einstein 'WAS RIGHT', but he did 'NOT QUITE' structure his original EPR argument into a winning form.
- 'Local Objective Reality' is the CORRECT MODEL!
- 'Instantaneous non-local entanglement' is WRONG!
- 'Instantaneous non-local causality' is WRONG!
- The 'Heisenberg Uncertainty Principle', when taken to imply a lack of 'physical existence' and 'physical persistence', is WRONG!

- 'Quantum Waves' being interpreted as non-physical 'probability waves' is WRONG!
- The idea of 'ghostly parallel quantum universes' is WRONG!

In lieu of the 'Standard Copenhagen Model of Quantum Mechanics', we can look forward to a new Quantum Mechanics based on *Objective Physicality.*

This all raises serious questions about the efficacy for proposed future 'Quantum Technologies', such as 'Quantum Computing'. Such technologies would utilize the properties of 'parallel quantum existences' and 'instantaneous non-local quantum entanglement'. However, if this parallel existence and instantaneous entanglement are a myth, how can we expect to develop its derivative technologies? In light of the 'New Quantum Reality', the outlook for such 'Quantum Technologies' does not appear promising. Accordingly, we should be prepared to question any proposed 'Quantum Technology' that appears to be an extension of the 'hoax' created by 'Quantum Mythology'.

Chapter Article 1.10

The Formation of Cosmutons

My revelation of *Unified Physicality* can be briefly stated as follows:

> The physical medium of three-dimensional space exists and is now to be called the *Cosmic Continuum.* The *Cosmic Continuum* is the sole physical substance of the universe. The *Cosmic Continuum* can be distorted in a dynamic way, while governed by a *Principle of Local Causality.* Distortions of the *Cosmic Continuum* form all elementary particles and all force fields. Self-supporting whirlwind 'solitonic' waves within the *Cosmic Continuum* form what we perceive as all the elementary particles. These elementary particles are to be called *Cosmutons*, an acronym repre-

senting "*COSmic Continuum MUtable SoliTONs*". (The accent is on the first syllable, ***Cos'-mu-ton***).

The *Cosmic Continuum* is like a compressible elastic solid. It has the composite qualities of 'jello-sponge-steel' ('jello' because it wiggles, 'sponge' because it is compressible, and 'steel' because it has resilient strength). We should not expect to fully comprehend this substance called the *Cosmic Continuum*. I postulate its existence and infer its *Objective Physical Properties* so that it has physical qualities sufficient to account for all the natural phenomena we observe.

A 'soliton' is a wave phenomenon generally known within science, whereby a single solitary wave can persist indefinitely without dispersion or decay. Conditions of nonlinear dynamics must be just right for it to form a soliton. When solitons do form, they possess both the quality of a stand-alone entity with an underlying structure of a wave. *Cosmutons* thus possess the innate quality of 'wave-particle duality'.

One feature of a *Cosmuton* is that it forms what can be considered to be a *Field String*. The surface boundary between an inner *Core Region* of vibration and an outer surrounding *Aura Region* of vibration constitutes a vibrating surface called the *Interseptum* or just *Septum*. The vibrating surface wave of the *Septum* is the physical incarnation of a *Field String*. These *Field Strings* are a wave vibration that realize the mathematical characteristics of 'extra-dimensional' String Theory, but do so within just the three spacial dimensions of the physical medium of space. This model of *Field String Formation* brings String Theory back into the realm of three-dimensional *Physical Reality*.

So how do we resolve the issue of the 'extra-dimensionality' of String Theory? Consider the variables for a donut shaped toroidal wave vortex of an *Electron's Septum*, which I also call a *Field String*. The surface of this donut can be described by 10 toroidal parameters. Two variables locate a relative position on the toroidal surface; two variables set the size of the major and minor toroidal radii; two variables orient the primary toroidal axis; three variables locate the vortex in global space; and a last variable evolves time. That adds up to 10 dynamic variables.

Thus, three-dimensional dynamics in the *Aura* and *Core* of a *Cosmutonic Electron* can produce a 10 variable dynamic within the two-dimensional *Electron Septum*. We can thereby reassert the preeminence of three dimensionality - NOT a 10-dimensional hypersurface in an 11-dimension hyperspace as touted in String Theory. Thus, the mathematical vagaries of 'extra-dimensional' String Theory can find a physical incarnation in the form of three-dimensional *Cosmutons*.

The *Core* and *Aura* waves of a *Cosmuton* must be in a state of synchronization at their interface, so that the *Field String* at the *Septum* is in a state of harmonic resonance. The various modes of harmonic resonance of the *Field String* constitute respective states of quantized energy. Rest-mass *Cosmutons* become distorted by absolute motion and gravitational fields to manifest the characteristics of Relativity. Absolute motion and gravitational fields alter the feedback pressures on rest-mass *Cosmutons* to smoothly alter their quantum states. This shifting of rest-mass *Cosmutonic* energy-states constitutes the *Physical Reality* underlying Relativity and the unification of Relativity and Quantum Mechanics.

A particular resonance mode of a rest-mass *Cosmuton* can 'mutate' to another state of resonance. This is not just a smooth shift of energy, but is a different configuration of vibration that alters the number and/or arrangement of the vibration nodes. In doing so, it must normally absorb or emit another *Cosmuton*, like an electron absorbing or emitting a photon. *Cosmutons* can thus interact and 'mutate' to manifest the characteristics of the many types of particles and particle interactions that are observed. The formation and 'mutation' of a *Cosmutonic Field String* with wave resonant harmonic modes constitutes the *Objective Physical Reality* underlying Quantum Mechanics.

Two different rest-mass particles can have complementing properties, such as positive and negative electric charge. When two particles are complementary in almost all ways possible, they can be to each other what is called 'anti-matter'. If they come into contact with each other they will combine to cancel out their complementary properties and produce non-rest-mass particles. They appear to annihilate one another, but their combined properties are assumed by their by-product particles. Thus 'matter' and 'anti-matter' have the same underlying composition, as they are both formed as wave disturbances of the *Cosmic Continuum*.

The exterior standing pressure wave of the *Aura* (which can also be called a sub-quantum wave) is a necessary feature of *Cosmutonic Formation*.

- The phasing of the exterior (in and out) sub-quantum *Auric Wave* must match the natural resonances of the *Cosmutonic Field String* (both the twisting and the looping of the vortex wave) and interior (in and out) of the *Cosmutonic Core*.
- The standing pressure wave of the *Aura* consists of outgoing and an incoming wave, and the incoming wave provides the critical feedback pressure necessary to hold the solitonic formation together.
- The size and shape of the soliton adjusts itself to maintain harmonic balance between the internal and external wave vibrations. This smooth variability is the feature that unifies Relativity and Quantum Mechanics.
- When all resonances are matched in phase and amplitude, the sub-quantum standing pressure *Auric Wave* provides a net inward pull. This creates a compression density gradient within the *Aura* (with density dropping off with distance away from the *Cosmuton*).
- In symbiotic support, this density gradient enables the distributed reflection of the outward pressure wave to produce a returning inward pressure wave to form the sub-quantum standing-pressure wave.

The *Cosmutonic Aura* is thus a combination of a *Compression Gradient* and a *Standing Pressure Wave* that symbiotically support each other. Together they provide the critical and necessary feedback-pressure required to sustain the very formation of a *Cosmuton*. The *Cosmutonic Aura* is able to form because the *Cosmic Continuum* behaves in a non-linear manner at the scaling sizes of *Cosmutonic Formation*.

Chapter Article 1.11

The Electromagnetic Buildup of Matter

Let us continue the story of the *Cosmutonic Aura* and see how it further plays its critically important role. For purposes of providing a brief introductory outline, with more details to be discussed in the *Full-Book Edition*, please consider the following highlights. The *Cosmutonic Aura:*

- Supports the formation of elementary particles,
- Manifests the elementary forces without mediating particles,
- Contributes to an altered buildup of matter,
- Forms physical quantum waves that can physically interfere,
- Produces a version of *Local Quantum Entanglement.*

Auric Formation of Elementary Forces: The static and dynamic variations of compression within the *Cosmic Continuum* induce the elementary forces. The elementary particles form as unique waves within the *Cosmic Continuum*. Just as gravity bends the path of light, distortions of the *Cosmic Continuum* bend the paths of a particle's internal waves to manifest the effects of all the truly elementary forces. The exterior *Standing Pressure Wave* of the *Aura* thus induces the elemental forces of gravity, electromagnetism, and nuclear binding.

- The compression density gradient within the *Aura* constitutes a gravitational field.
- A non-symmetric sub-quantum *Auric Wave* constitutes an electric field.
- The direction of non-symmetric slope of the sub-quantum *Auric Wave* establishes the direction of the electric field.
- A laterally phased (sideways drifting) electric field *Auric Wave* constitutes a magnetic field.
- The strong and weak nuclear forces are just subtle variations in the near-field geometry of the electromagnetic field.

- The reaction of objects to *Cosmic Continuum* communicated forces is itself a reaction force. For gravity, this reaction force will be called the *Gravio-Reaction Force*.
- The out of phase condition of the outward and inward components of the standing compression wave produce a *Quantum Force*, which de Broglie called the 'Pilot Force'.
- All elementary forces upon rest-mass particles could be mathematically expressed with additional gamma factors within $E=mc^2$, but not here in the *Monograph*.

Like dinosaurs stuck in the tar pits, some people will be unable to extricate themselves from the idea of force mediating particles. 'Supposedly', gravitons beget gravity, gluons beget the strong nuclear force, weakons beget the weak nuclear force, and photons beget electric and magnetic forces. We have never directly seen any of these particles in action. We merely infer they must exist because we observe that the forces exist. Consider this test, which you can perform. 'Supposedly', photons mediate the force of magnetism. Take two magnets and measure the force between them at some nominal distance. Now interpose some non-magnetic non-transparent material between the two magnets without moving the magnets. The magnetic force remains unchanged. The non-transparent material (which by definition of non-transparency blocks photons) fails to block any of the magnetic force mediating photons. No one has ever observed or measured magnetic force mediating photons. They are just a concoction contrived to satisfy a flawed theory of force mediating particles.

Auric Contribution to the Buildup of Matter: The exterior *Standing Pressure Wave* of the *Aura* contributes prominently in the buildup of matter. I disagree with the underlying physicality of David Bergman's spinning electric charge ring model for the electron and proton, though of course that is what my *Cosmutonic Field String* model outwardly looks like. However, remember that my *Field String* is just the boundary between two regions of dynamic wave distortions within the *Cosmic Continuum*. Bergman's ring model is based upon the existence of charge as the elemental substance without any physical continuum. Nevertheless, I do like how Bergman (in conjunction with Charles Lucas Jr.) describes the buildup of matter. My description of the buildup of matter is similar, but obviously different because I invoke the *Standing Pressure Auric Wave* within the *Cosmic Continuum*. The *Aura Wave* can manifest de Broglie quantum waves and a locally restricted form of quantum entanglement, which the Bergman/Lucas model cannot do.

- It is well known that electrons and protons possess electric charge, but it is NOT well known that they are also little magnets. *Cosmutonic* charged particles are spinning vortices that create their own magnetism. While like charges experience electrostatic repulsion, their magnetic axes will twist into north and south pole alignment and then experience magnetic attraction. Electrostatic repulsion normally overpowers magnetic attraction, but up close, magnetic attraction can sometimes overpower electrostatic repulsion. This view greatly alters our perception of the buildup of matter.
(NOTE: Electrostatic force varies as $1/r^2$, far field magnetic force varies as $1/r^3$, and near field magnetic force varies as $1/r$. To the mathematically minded reader this tells an amazing story for like charges being brought closer together. It says that if the far field magnetic force approximation holds true until after it exceeds the electric force, then the particles of like charge can find a region of net attraction. Most importantly, this leads to a stable hover point so that like charges can become electromagnetically locked at a fixed separation.)
- With proper inducement, a large *Cosmutonic* electron vortex can slide into an encircling position around a smaller *Cosmutonic* proton vortex. The proton vortex then electromagnetically pulls the electron vortex in closer to form a small *Cosmutonic Neutron*. The net outward appearance of charge is zero as the two charges cancel one another, but the magnetic strengths add together. Thus, the neutron appears to be electrically neutral but with a strong magnetic field induced by apparently

counter-rotating opposite electric charges. The ***Fictitious*** 'Weak Nuclear Force' is actually just a near-field subtlety of the electromagnetic force that holds *Elementary Cosmutons* together to form a *Compound Cosmuton*, such as a neutron.

- Magnetic protons with the aide of magnetic neutrons can magnetically overpower the electrostatic repulsion of protons to form a *Cosmutonic Atomic Nucleus*. The magnetic attraction between two protons is barely strong enough to overpower electrostatic repulsion and so such bonds are unstable. However, given a mix of protons and neutrons, the neutrons add additional magnetic binding strength without adding any additional electrostatic repulsion and thus can form an atomic nucleus. The 'Strong Nuclear Force' was originally proposed to hold atomic nuclei together, but that was before it was discovered elementary particles possessed spin and magnetic dipole moments. The ***Fictitious*** 'Strong Nuclear Force' is actually just another near-field subtlety of the electromagnetic force that holds protons and neutrons together within a *Cosmutonic Atomic Nucleus*.
- Within an atom, electrons balance electric and magnetic forces to take up quasi-stable positions about a nucleus to form a *Cosmutonic Atom*. Nominally, the positive nucleus attracts negative electrons, but this is complemented by both electron/electron electrostatic repulsion and magnetic attraction. While electrons would normally repel one another, that effect is reduced by magnetic attraction among the electrons. The electromagnetic clustering of electrons alone is not stable, but with the added electromagnetic forces from the nucleus, the electrons can cluster to form a quasi-static *Cosmutonic Atom*. The ***Hypothetical*** 'Electron Orbit' and 'Electron Quantum Cloud' models of the atom could conceivably be replaced by a 'Quasi-Stationary' model. The electron *Auric Waves* could serve to mimic the cloud model.
- *Cosmutonic Atoms* then can build up *Cosmutonic Molecules*, much as currently touted within Physics and Chem-

istry. Overall, the standing pressure *Auric Waves* inherent in the formation of all *Rest-mass Cosmutons* form miniature electromagnetic potential wells that secure the quasi-stable positioning of particles as they cluster together. The clustering can form *Compound Cosmutons* (such as neutrons), *Cosmutonic Atomic Nuclei*, *Cosmutonic Atoms*, and *Cosmutonic Molecules*.

Auric Formation of Quantum Waves: The exterior standing pressure wave of the *Aura* is a sub-quantum wave that forms de Broglie matter waves.

- When a rest-mass *Cosmuton* moves, it alters the in-and-out timing of the standing sub-quantum *Auric Wave*. This altered timing produces altered phasing within the sub-quantum *Auric Wave*.
- The exterior standing wave thus becomes modulated to form a traveling wave envelope.
(NOTE: For example, an amplitude modulated AM radio wave modulates the radio wave - the electromagnetic sub-carrier wave. The radio wave is modulated by an envelope pattern that matches the sound wave - the signal we expect to eventually hear. The sound wave does not travel to us, but the radio wave with the sound wave information does travel to us through air and space.)
- In the case of quantum waves, the sub-quantum *Auric Wave* becomes modulated. That is why I call the *Auric Wave* a sub-quantum wave. It is the sub-carrier for another superimposed quantum wave signal.
- This wave envelope of the sub-quantum *Auric Wave* constitutes the de Broglie matter wave, which can interfere to produce quantum interference via the *Quantum Force*.
- The wave envelope of the sub-quantum *Auric Wave* is the physical incarnation of the de Broglie wave, which has been misinterpreted as being a 'quantum probability wave'. This quantum wave is NOT a 'probability wave'. It is a 'physical wave'. More precisely, it is the envelope of a physical wave.

Auric Contribution to Entanglement: The exterior standing pressure wave of the *Aura* contributes to a subtle restricted version of quantum entanglement.

- When a *Cosmutonic* photon passes from one *Cosmutonic* electron to another (or from one *Cosmutonic* proton to another), it passes thermal energy and thus establishes a '*Thermal Entanglement*' for *Cosmutonic* particles that are otherwise separated.
- When *Cosmutonic* photons pass between *Cosmutonic* particles that are relatively close to one another, the emitted photons never fully form before they begin to be absorbed. This form of '*Thermal Entanglement*' can be described as '*Virtual Entanglement*' and the incompletely formed photons can be described as being '*Virtual Photons*'. (This is a completely different sense then what is currently ascribed to the term '*Virtual Photons*', which fallaciously touts them as being the never observed carriers of electomagnetic force.)
- If the sub-quantum *Auric Wave* of one electron stimulates emission of a photon by a second electron, and if this emitted photon is absorbed by the first electron, we have '*Sympathetic Entanglement*'. It is as if the first electron reached out and extracted the photon from the second electron via this '*Sympathetic Entanglement*'. (This '*Sympathetic Entanglement*' is only relatively local in its reach via the *Auric Wave* and is not long range nor instantaneous such as the entanglements touted within 'Extreme Quantum Weirdness'.)

In conclusion, distortions of the *Cosmic Continuum* constitute a *Physical Field.* Dynamic disturbances of this *Physical Field* can form *Cosmutons* with their unique characteristic of a *Cosmutonic Field String.* We can identify the *Cosmutonic Core* as being an elementary particle and the *Cosmutonic Auras* as forming a sub-quantum exterior wave. The sub-quantum wave forms the elementary force fields and can also be modulated to form de Broglie quantum waves. Thereafter, *Cosmutons* can then come into quasi-stable positions with respect to one another to form the buildup of matter.

Chapter Article 1.12

Restoring Objective Physical Reality

The eye of the beholder can recognize beauty, but the eye of our mind is a powerful resource able to discern truth and reality. However, biases can so confuse our ability to discern, that we should always be prepared to question and to reevaluate. In questioning and reevaluating Modern Physics, I now conclude we have been led astray. We have been improperly led to reject *Objective Physical Reality*, and this has been as wrong as wrong can be!

Einstein tried to envision a universe based upon *Objective Physicality*. However, the 'obfuscating mathematical' representations of natural phenomena by Modern Physics have served to hide the 'actual physicalities' underlying physics. I claim the observations of all natural phenomena can indeed be based upon an underlying *Common Sense Physicality*. In presenting the plausibility of an underlying physicality, the corrected mathematical formulations of Modern Physics can then be re-attached to *Objective Physical Realities*.

The purpose of this *Monograph* and the *Full-Book Edition* is to present the story of *Unified Physicality* so that average readers, as well as physicists, can readily understand my vision of *Common Sense Physicality*. In so doing, I see a five-fold story line:

- To reveal the one *Objective Physicality* underlying all natural phenomena.
- To physically justify those abstractions of Modern Physics which are valid.
- To debunk those aspects of Modern Physics which are outright fallacies.

- To re-establish the preeminence of *Objective Common Sense Physical Reality.*
- To present *Unified Physical Field Theory* as a credible theory without the burden of unnecessary mathematical baggage.

In the late 1600s, the incomparable Sir Isaac Newton set the standard of discerning *Physical Causality* as the foundation of science. However, for attempting to describe an underlying physicality of forces, Newton was rebuked by his peers. In the early 1900s, Albert Einstein valiantly defended the standard of *Physical Causality* against the unworthy assault by Quantum Physicists. Einstein appeared to lose his battle against those proponents of non-physicalities, and he also was rebuked by his peers. I now take up the mission of *Physical Causality* championed by both Newton and Einstein and call for Modern Physicists to renounce their 'sophistry of anti-physicality'. In making this plea, let there be no mistake. The confirmed observations of Modern Physicists deserve proper consideration. In the *Full-Book Edition*, I seek to credibly explain the confirmed subtleties of Modern Physics, and I will do so based upon my vision of *Objective Common Sense Physicality.*

Einstein's presentation of Relativity has created serious difficulties, but these can be overcome without too much intellectual pain. I believe Einstein would welcome my suggested clarifications. However, Niels Bohr's Standard Model of Quantum Mechanics, which is known as the 'Copenhagen Convention', is unbelievably 'anti-physical'. This model of Quantum Mechanics is the antithesis of physicality. I believe Einstein understood that if the Copenhagen Model were allowed to stand, no sound theory of *Physical Unification* could ever be developed! I agree with this sentiment! However, sentiment is not enough. To restore *Objective Physical Causality* as the proper basis of science, we MUST expose the fallacy of the Copenhagen Model of Quantum Mechanics. I am prepared to do this!

I have presented two revolutionary examples to refute the current formulation of Modern Physics and to support my assertion for a *Unified Physical Field Theory.* By 'Perfecting

$E=mc^2$' and 'Debunking Quantum Weirdness', I believe we forever change the landscape of physics. We expose the fallacy of the mathematical voyeurism that has drawn Modern Physics into the 'sophistry of anti-physicality'. We thus set the stage for restoring *Objective Physical Reality.* Let us briefly summarize the main highlights of my two dramatic examples.

- ***Perfecting E=mc²:***

  $$E = mc^2 = \cancel{m_o c^2 \gamma} = m_o c_o^{\,2} \gamma_M^{\,1} \gamma_G^{\,-1}$$

  Gravitational Redshift is induced by both *Gravitational Time Dilation* and *Gravitational Rest-Energy Diminishment.* We thereby conclude the following: The *Cosmic Continuum* is the physical medium of space that fills and forms our three-dimensional universe. Gravity does not curve space but does make it non-Euclidean. 'Blackholes' do not exist but *Greyholes* do exist. Quasars are *Greyholes* that reside near and interact with galaxies. Dark Matter does not exist but a *Gravio-Reaction Force* does exist to produce the anomalous rotations of galaxies.
- ***Debunking Quantum Weirdness:*** *Bell's Inequality* for photons is incorrectly formulated, and thus, *EPR Experiments* are misinterpreted. We thereby conclude the following: ***Non-Physical*** 'quantum probability waves', ***Ghostly*** 'parallel universes', ***Instantaneous*** 'non-local entanglement', and ***Instantaneous*** 'non-local causality' are ***Invalid*** 'anti-physical' concepts. Let us return to *Common Sense Physicality* that dwells within three-dimensional space without any ***Extra*** 'dimensions' or ***Infinite*** 'singularities'. Let us be guided by a *Principle of Local Physical Causality.*

Concerning these two key issues of *Gravitational Redshift* and 'Extreme Quantum Weirdness', Einstein 'WAS RIGHT' but 'NOT QUITE' complete. My vision of *Objective Physical Reality* completes what Einstein began and thus leads to the assertion of four new fundamental truths:

- 'Relativity' and 'curved spacetime' are compatible with a three-dimensional physical medium of space, and this medium of space possesses an *Objective Physical Existence*! (Relativity's fallacious introduction of an 'extra dimension' or two has been an 'anti-physical' misadventure. The rejection of a physical medium of space, though not 'anti-physical', has nonetheless been an horrendous mistake.)
- 'Quantum Mechanics' can be re-structured to be compatible with both a *Principle of Local Causality* and with the idea of a *Localized Objective Physical Existence*! (The introduction of 'non-physical probability waves' has been quite bizarre and 'anti-physical'. The fallacious advancement of 'instantaneous non-local causality' and 'ghostly parallel universes' has been extremely 'anti-physical'. The preposterous claim that 'uncertainty of measurement' implies 'uncertainty of physical existence' has been the notorious 'anti-physical' root of 'Extreme Quantum Weirdness'.)
- 'String Theory' (including its most touted variation 'M-Theory') can be re-structured to be compatible with a set of physical variables inherent with an *Objective Physicality* that dwells within a physical medium of three-dimensional space! (The unrestrained mathematical voyeurism into multiple 'extra-dimensional' Strings has been an 'anti-physical' miscue induced by the misuse of mathematical semantics.)
- 'Particle Theories' (including quarks, Higgs Bosons, and Strings) can be re-structured to be compatible with the idea a physical medium of three-dimensional space! The elementary particles form as dynamic distortions of the physical medium of three-dimensional space! The medium is to be called the *Cosmic Continuum*, and the elementary particle formations are to be called *Cosmutons*. Distortions of the *Cosmic Continuum* also communicate the elementary forces.

Einstein was trying to develop a 'Unified Field Theory' via an 'Abstract Field'. In contrast, I employ a '*Physical Field'* within the physical medium of space to formulate my '*Unified Physical Field Theory*'. I call this physical medium of three-dimensional space the *Cosmic Continuum*. I claim that unique wave disturbances within the *Cosmic Continuum* form all the elementary particles, and I call these formations *Cosmutons*. I further claim that the distortions of the *Cosmic Continuum* surrounding the heart of these *Cosmutons* form all the truly elementary forces. This then constitutes the specific underlying *Common Sense Physicality* for all natural phenomena in today's single universe, which consists of just three spacial dimensions.

Newton and Einstein (the two greatest physicists of all time) were both rejected and insulted for their perceptions of *Objective Physical Reality*. In direct contrast, I salute them for their bold visions of physicality and their bold expression of their visions. For their visions of *Objective Physical Reality*, Newton and Einstein deserve our renewed admiration. Having presented the Universal Law of Gravitation, Newton is often misrepresented as advocating instantaneous 'action-at-a-distance', which he explicitly rebuked. In seeking to develop mathematical representations for an 'Abstract Field', Einstein is often misinterpreted as abandoning *Objective Physicality* and rejecting a medium of space, which he did not do. To avoid misunderstanding, I openly seek to develop physical perspectives about a '*Physical Field*'. Hence, I pursue a '*Unified Physical Field Theory*' based upon *Objective Common Sense Physical Reality*.

Chapter 1 Postscript: The roles of mathematics and *Objective Physicality* within physics needs to be put into proper perspective. We need to restore a proper sense of *Objective Physicality.*

Einstein stated:

***"All physical theories,
their mathematical expressions notwithstanding,
ought to lend themselves to so simple a description
that even a child could understand them.***

I agree and add this complementing thought:

*Mathematics devoid of Objective Physicality
is just Pure Mathematics
(not bad, but Not Physics),*

*While Mathematics depicting Actual Physicality
is Real and Proper Physics.*

*Mathematical-Physics touting Anti-Physicalities
is Scientific Mysticism,
(a Sophistry of Anti-Physicality),*

*While Physics restoring Objective Physicality
is a return to Scientific Sanity.*

Monograph Postscript: For further discussions, details, justifications, rationales, and explanations, read the *Full-Book Edition*. Therein, I shall further explore the ideas of the scientific pioneers of ancient times, of Newton's Era of Mechanics, of Maxwell's Era of Electromagnetism, and of Einstein's Era of Modern Physics. I shall more systematically expose the fallacies of Modern Physics. I shall more precisely present the new vision of Unified Physicality. In various appendices, I shall present (if you are interested) more detailed support material.

This story of *Unified Physicality* is unique and revolutionary. You should feel both troubled by the past transgressions of Modern Physics and encouraged to consider a vision for a new era of science - '*The Era of Unified Physics*'. This story of *Unified Physicality* may also give you pause to review the validity of other troubling mantras of our modern society.

This Ending is Just the Beginning!

Once we open a crack in the wall of Anti-Physicality,

A flood of Common Sense Ideas will burst through;

Ideas pouring forth like a Deluge;

Ideas sweeping away Foolishness;

Ideas clearing the way for a New Enlightenment!

Encore - One Last Thought

This last insight is just too good to delay until the *Full-Book Edition*, as I originally planned. This last insight culminates with a most curious question.

To begin, let us ask this. Can two photons interact? Yes they can. If they have enough energy and interact strongly, they annihilate each other and produce rest-mass particles. This is an observed and proven fact. Two gamma ray photons annihilate each other to produce an electron-positron pair. In reverse, an electron and a positron (which are a 'matter' and 'anti-matter' pair) will annihilate each other and produce a pair of gamma ray photons. They are all made out of the same substance. They are all formed as distortions of the *Cosmic Continuum*.

Could two photons interact very weakly? Could two photons lose just a little energy, and otherwise continue their existence, but produce a third very-low energy photon? What if that did happen, but only rarely? What would be the result? We might not be able to see the effect for earth based experiments, because the times of photon travel would be too short to observe a very rare event. What if we look at the light coming from distant stars? Could we conceivably see any indication of very-weak and rare photon interactions?

The very-weak and rare interaction of long lived cosmic photons would diminish their energies. They would appear red-shifted, and the redshift would be proportional to distance. Do we see any such redshift proportional with distance? Yes we do. We see the Hubble Redshift, which is also called the Cosmological Redshift. This has been considered justification for the idea of an expanding universe, the big bang, and dark energy. Could the very-weak and rare interaction of cosmic photons contribute in whole or in part to the Hubble Redshift? Why not?

The very-weak and rare interaction of long lived cosmic photons would produce a sea of very low energy by-product photons. Do we see any such sea of very low energy photons? Yes we do. We see the Cosmic Microwave Background Radiation, which has been considered a confirmation of the big bang. Could the very-weak and rare interaction of cosmic photons contribute in whole or in part to production of the Cosmic Microwave Background Radiation? Why Not?

Historically, the possibility of photons losing energy has been described as Tired Light. It was dismissed as not viable because it was assumed that Tired Light would become diffuse, and we observe that star light is not diffuse. However, the dismissal of Tired Light may have been premature.

Very-weakly interacting *Cosmutonic Photons* need not produce diffuse light. *Cosmutonic Photons* possess a distributed physical existence, yet normally they do not appear to interact when they pass through one another. We can imagine that *Cosmutonic Photons* act like little gyroscopes and recover their stability and their course of motion after they encounter a non-altering interaction. When *Cosmutonic Photons* experience a very-weak interaction, the same principle of trajectory recovery could apply, so that they do not need to form diffuse light. Thus, Tired Light could conceivably be restored as a viable concept.

Recall also, that *Greyhole-Quasars* should be able to formulate heavier elements, which some previously perceived as only being produced by the big bang.

Dark energy and the big bang are both justified primarily because of our prior understandings about Cosmological Redshift being induced by an expanding universe. And the big bang model is supposedly confirmed by the existence of the Cosmic Microwave Background Radiation and the existence of the heavier elements. However, if a *Cosmutonic Model of Photons* can produce the Cosmological Redshift and the Cosmic Microwave Background Radiation, and if *Greyhole-Quasars* can formulate the heavier elements, what happens to the efficacy of dark energy and the big bang?

The fabric of space, hereafter to be called the *Cosmic Continuum,* is now to be understood to be the elemental substance underlying all the natural phenomena that we observe. How has it evolved, if any, over the epochs of time? If it has changed, how would we know? What would we look for? Now we are getting into far reaching speculations that exceed the intended scope of this *Monograph.*

Having stated all of the above, I believe we should prepare ourselves to re-evaluate the 'HYPOTHESES' of an expanding universe, dark energy, and the big bang! The question I wish to leave you with is this. How much, if any, might Big Bang Cosmology change? Let the discussion begin.

End of Monograph